THE
DEEP LEARNING WITH KERAS
WORKSHOP

Learn how to define and train neural network models with just a few lines of code

Matthew Moocarme, Mahla Abdolahnejad, and Ritesh Bhagwat

THE DEEP LEARNING WITH KERAS WORKSHOP

Authors: Matthew Moocarme, Mahla Abdolahnejad, and Ritesh Bhagwat

Reviewers: Vikraman Karunanidhi, Asheesh Mehta, Bernard Ong, and Anuj Shah

Managing Editor: Bhavesh Bangera

Acquisitions Editors: Kunal Sawant, Archie Vankar, and Karan Wadekar

Production Editor: Roshan Kawale

Editorial Board: Megan Carlisle, Samuel Christa, Mahesh Dhyani, Heather Gopsill, Manasa Kumar, Alex Mazonowicz, Monesh Mirpuri, Bridget Neale, Dominic Pereira, Shiny Poojary, Abhishek Rane, Brendan Rodrigues, Erol Staveley, Ankita Thakur, Nitesh Thakur, and Jonathan Wray

First published: July 2020

Production reference: 2230221

ISBN: 978-1-80056-296-7

Published by Packt Publishing Ltd.

Livery Place, 35 Livery Street

Birmingham B3 2PB, UK

WHY LEARN WITH A PACKT WORKSHOP?

LEARN BY DOING

Packt Workshops are built around the idea that the best way to learn something new is by getting hands-on experience. We know that learning a language or technology isn't just an academic pursuit. It's a journey towards the effective use of a new tool— whether that's to kickstart your career, automate repetitive tasks, or just build some cool stuff.

That's why Workshops are designed to get you writing code from the very beginning. You'll start fairly small—learning how to implement some basic functionality—but once you've completed that, you'll have the confidence and understanding to move onto something slightly more advanced.

As you work through each chapter, you'll build your understanding in a coherent, logical way, adding new skills to your toolkit and working on increasingly complex and challenging problems.

CONTEXT IS KEY

All new concepts are introduced in the context of realistic use-cases, and then demonstrated practically with guided exercises. At the end of each chapter, you'll find an activity that challenges you to draw together what you've learned and apply your new skills to solve a problem or build something new.

We believe this is the most effective way of building your understanding and confidence. Experiencing real applications of the code will help you get used to the syntax and see how the tools and techniques are applied in real projects.

BUILD REAL-WORLD UNDERSTANDING

Of course, you do need some theory. But unlike many tutorials, which force you to wade through pages and pages of dry technical explanations and assume too much prior knowledge, Workshops only tell you what you actually need to know to be able to get started making things. Explanations are clear, simple, and to-the-point. So you don't need to worry about how everything works under the hood; you can just get on and use it.

Written by industry professionals, you'll see how concepts are relevant to real-world work, helping to get you beyond "Hello, world!" and build relevant, productive skills. Whether you're studying web development, data science, or a core programming language, you'll start to think like a problem solver and build your understanding and confidence through contextual, targeted practice.

ENJOY THE JOURNEY

Learning something new is a journey from where you are now to where you want to be, and this Workshop is just a vehicle to get you there. We hope that you find it to be a productive and enjoyable learning experience.

Packt has a wide range of different Workshops available, covering the following topic areas:

- Programming languages

- Web development

- Data science, machine learning, and artificial intelligence

- Containers

Once you've worked your way through this Workshop, why not continue your journey with another? You can find the full range online at http://packt.live/2MNkuyl.

If you could leave us a review while you're there, that would be great. We value all feedback. It helps us to continually improve and make better books for our readers, and also helps prospective customers make an informed decision about their purchase.

Thank you,
The Packt Workshop Team

Table of Contents

Chapter 6: Model Evaluation 213

Chapter 7: Computer Vision with Convolutional Neural Networks 249

PREFACE

ABOUT THE BOOK

New experiences can be intimidating, but not this one! This beginner's guide to deep learning is here to help you explore deep learning from scratch with Keras, and be on your way to training your first ever neural networks.

What sets Keras apart from other deep learning frameworks is its simplicity. With over two hundred thousand users, Keras has a stronger adoption in industry and the research community than any other deep learning framework.

The Deep Learning with Keras Workshop starts by introducing you to the fundamental concepts of machine learning using the scikit-learn package. After learning how to perform the linear transformations that are necessary for building neural networks, you'll build your first neural network with the Keras library. As you advance, you'll learn how to build multi-layer neural networks and recognize when your model is underfitting or overfitting to the training data. With the help of practical exercises, you'll learn to use cross-validation techniques to evaluate your models and then choose the optimal hyperparameters to fine tune their performance. Finally, you'll explore recurrent neural networks and learn how to train them to predict values in sequential data.

By the end of this book, you'll have developed the skills you need to confidently train your own neural network models.

AUDIENCE

If you know the basics of data science and machine learning and want to get started with advanced machine learning technologies like artificial neural networks and deep learning, then this is the book for you. To grasp the concepts explained in this deep learning book more effectively, prior experience in Python programming and some familiarity with statistics and logistic regression are a must.

ABOUT THE CHAPTERS

Chapter 1, Introduction to Machine Learning with Keras, will introduce you to the fundamental concepts of machine learning by using the scikit-learn package. You will learn how to present data for model building, then train a logistic regression model using a real-world dataset.

Chapter 2, Machine Learning versus Deep Learning, will present the difference between traditional machine learning algorithms and deep learning algorithms. You will learn the linear transformations necessary for building neural networks and build your first neural network with the Keras library.

Chapter 3, Deep Learning with Keras, will extend your knowledge of neural network building. You will learn how to build multi-layer neural networks and recognize when your model is underfitting or overfitting to the training data.

Chapter 4, Evaluating Your Model with Cross-Validation Using Keras Wrappers, will teach you how to use Keras wrappers with scikit-learn to incorporate Keras models into a scikit-learn workflow. You will apply cross-validation to evaluate your models and use this technique to choose the optimal hyperparameters.

Chapter 5, Improving Model Accuracy, will introduce various regularization techniques to prevent your models from overfitting to the training data. You will learn different methods to search for the optimal hyperparameters that result in the highest model accuracy.

Chapter 6, Model Evaluation, will demonstrate a variety of methods to evaluate your models. Beyond accuracy, you will learn more model evaluation metrics including sensitivity, specificity, precision, false positive rate, ROC curves, and AUC scores to understand how well your models perform.

Chapter 7, Computer Vision with Convolutional Neural Networks, will introduce you to building image classifiers with convolutional neural networks. You will learn about all the components that comprise the architecture of convolutional neural networks and then build image processing applications to classify images.

Chapter 8, Transfer Learning and Pre-Trained Models, will introduce you to the concept of transferring the learning from one model to solve for other applications. You will achieve this by using different pre-trained models and modifying them slightly to fit different applications.

Chapter 9, Sequential Modeling with Recurrent Neural Networks, will teach you how to build models with sequential data. You will learn the architecture of recurrent neural networks and how to train them to predict the succeeding values from sequential data. You will test your knowledge by predicting the future values of various stock prices.

CONVENTIONS

Code words in text, database table names, folder names, filenames, file extensions, path names, dummy URLs, user input, and Twitter handles are shown as follows:

"**sklearn** has a class called **train_test_split**, which provides the functionality for splitting data."

Words that you see on the screen, for example, in menus or dialog boxes, also appear in the same format.

A block of code is set as follows:

```
# import libraries
import pandas as pd
from sklearn.model_selection import train_test_split
```

New terms and important words are shown like this:

"A dictionary contains multiple elements, like a **list**, but each element is organized as a key-value pair."

CODE PRESENTATION

Lines of code that span multiple lines are split using a backslash (\ ). When the code is executed, Python will ignore the backslash, and treat the code on the next line as a direct continuation of the current line.

For example:

```
history = model.fit(X, y, epochs=100, batch_size=5, verbose=1, \
                    validation_split=0.2, shuffle=False)
```

Comments are added into code to help explain specific bits of logic. Single-line comments are denoted using the # symbol, as follows:

```
# Print the sizes of the dataset
print("Number of Examples in the Dataset = ", X.shape[0])
print("Number of Features for each example = ", X.shape[1])
```

Multi-line comments are enclosed by triple quotes, as shown below:

```
"""
Define a seed for the random number generator to ensure the
result will be reproducible
"""
seed = 1
np.random.seed(seed)
random.set_seed(seed)
```

SETTING UP YOUR ENVIRONMENT

Before we explore the book in detail, we need to set up specific software and tools. In the following section, we shall see how to do that.

INSTALLING ANACONDA

For this course, we will use Anaconda; a Python distribution which comes with an in-built package manager and pre-installed packages frequently used for machine learning and scientific computing,

To install Anaconda, find your desired version (Windows, macOS, or Linux) on the official installation page at https://docs.anaconda.com/anaconda/install/. Follow the appropriate installation instructions for your operating system.

Once Anaconda is installed, you can interact with it via the Anaconda Navigator or Anaconda Prompt. There are instructions on how to use these at https://docs.anaconda.com/anaconda/user-guide/getting-started/.

To verify the correct installation, you can execute the **anaconda-navigator** command on CMD / Terminal. If the installation is correct, this will open up the Anaconda Navigator.

INSTALLING LIBRARIES

`pip` comes pre-installed with Anaconda. Once Anaconda is installed on your machine, all the required libraries can be installed using `pip`, for example, `pip install numpy`. Alternatively, you can install all the required libraries using `pip install -r requirements.txt`. You can find the `requirements.txt` file at https://packt.live/3hhZ2v9.

The exercises and activities will be executed in Jupyter Notebooks. Jupyter is a Python library and can be installed in the same way as the other Python libraries – that is, with `pip install jupyter`, but fortunately, it comes pre-installed with Anaconda.

RUNNING JUPYTER NOTEBOOK

You can start Jupyter via the appropriate link in the Anaconda Navigator, or by executing the command `jupyter notebook` in Anaconda Prompt / CMD / Terminal.

Jupyter will open up in your browser, where you will then be able to navigate to your working directory and create, edit and run your code files.

ACCESSING THE CODE FILES

You can find the complete code files of this book at https://packt.live/2OL5E9t. You can also run many activities and exercises directly in your web browser by using the interactive lab environment at https://packt.live/2CXyFLS.

We've tried to support interactive versions of all activities and exercises, but we recommend a local installation as well for instances where this support isn't available.

The high-quality color images used in the book can found at https://packt.live/2u9Tno4.

If you have any issues or questions about installation, please email us at `workshops@packt.com`.

1

INTRODUCTION TO MACHINE LEARNING WITH KERAS

OVERVIEW

This chapter introduces machine learning with Python. We will use real-life datasets to demonstrate the basics of machine learning, which include preprocessing data for machine learning models and building a classification model using the logistic regression model with scikit-learn. We will then advance our model-building skills by incorporating regularization into our models and evaluating their performance with model evaluation metrics. By the end of this chapter, you will be able to confidently create models to solve classification tasks using the scikit-learn library in Python and evaluate the performance of those models effectively.

INTRODUCTION

Machine learning is the science of utilizing machines to emulate human tasks and to have the machine improve its performance of that task over time. By feeding machines data in the form of observations of real-world events, they can develop patterns and relationships that will optimize an objective function, such as the accuracy of a binary classification task or the error in a regression task.

In general, the usefulness of machine learning is in the machine's ability to learn highly complex and non-linear relationships in large datasets and to replicate the results of that learning many times. One branch of machine learning algorithms has shown a lot of promise in learning highly complex and non-linear relationships associated with large, often unstructured datasets such as images, audio, and text data—**Artificial Neural Networks (ANNs)**. ANNs, however, can be complicated to program, train, and evaluate, and this can be intimidating for beginners in the field. Keras is a Python library that presents a facile introduction to building, training, and evaluating ANNs that is incredibly useful to those studying machine learning.

Take, for example, the classification of a dataset of pictures of either dogs or cats into classes of their respective type. For a human, this is simple, and the accuracy would likely be very high. However, it may take around a second to categorize each picture and scaling the task can only be achieved by increasing the number of humans, which may not be feasible. While it may be difficult, though certainly not impossible, for machines to reach the same level of accuracy as humans for this task, machines can classify many images per second, and scaling can be easily done by increasing the processing power of a single machine or making the algorithm more efficient:

Image   Label

? ? ? ?

Figure 1.1: The classification of images as either dog or cat is a simple task
for humans, but quite difficult for machines

While the trivial task of classifying dogs and cats may be simple for us humans, the same principles that are used to create a machine learning model that classifies dogs and cats can be applied to other classification tasks that humans may struggle with. An example of this is identifying tumors in Magnetic Resonance Images (MRIs). For humans, this task requires a medical professional with years of experience, whereas a machine may only need a dataset of labeled images. The following image shows MRI images of the brain, some of which include tumors:

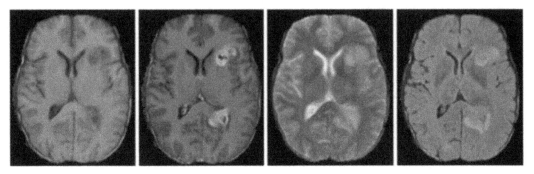

Figure 1.2: A non-trivial classification task for humans – MRIs of brains, some of which include the presence of tumors

DATA REPRESENTATION

We build models so that we can learn something about the data we are training on and about the relationships between the features of the dataset. This learning can inform us when we encounter new observations. However, we must realize that the observations we interact within the real world and the format of the data that's needed to train machine learning models are very different. Working with text data is a prime example of this. When we read the text, we are able to understand each word and apply the context that's given by each word in relation to the surrounding words—not a trivial task. However, machines are unable to interpret this contextual information. Unless it is specifically encoded, they have no idea how to convert text into something that can be a numerical input. Therefore, we must represent the data appropriately, often by converting non-numerical data types—for example, converting text, dates, and categorical variables into numerical ones.

TABLES OF DATA

Much of the data that's fed into machine learning problems is two-dimensional and can be represented as rows or columns. Images are a good example of a dataset that may be three-or even four-dimensional. The shape of each image will be two-dimensional (a height and a width), the number of images together will add a third dimension, and a color channel (red, green, and blue) will add a fourth:

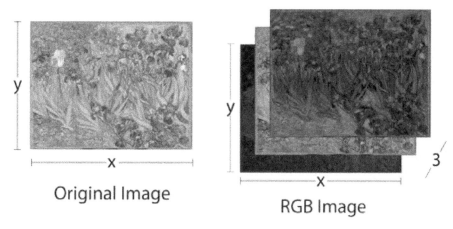

Original Image   RGB Image

Figure 1.3: A color image and its representation as red, green, and blue images

The following figure shows a few rows from a dataset that has been taken from the UCI repository, which documents the online session activity of various users of a shopping website. The columns of the dataset represent various attributes of the session activity and general attributes of the page, while the rows represent the various sessions, all corresponding to different users. The column named **Revenue** represents whether the user ended the session by purchasing products from the website.

> **NOTE**
>
> The dataset that documents the online session activity of various users of a shopping website can be found here: https://packt.live/39rdA7S.

One objective of analyzing the dataset could be to try and use the information given to predict whether a given user will purchase any products from the website. We can then check whether we were correct by comparing our predictions to the column named **Revenue**. The long-term benefit of this is that we could then use our model to identify important attributes of a session or web page that may predict purchase intent:

Related_Duration	BounceRates	ExitRates	PageValues	SpecialDay	Month	OperatingSystems	Browser	Region	TrafficType	VisitorType	Weekend	Revenue
0.000000	0.200000	0.200000	0.0	0.0	Feb	1	1	1	1	Returning_Visitor	False	False
64.000000	0.000000	0.100000	0.0	0.0	Feb	2	2	1	2	Returning_Visitor	False	False
0.000000	0.200000	0.200000	0.0	0.0	Feb	4	1	9	3	Returning_Visitor	False	False
2.666667	0.050000	0.140000	0.0	0.0	Feb	3	2	2	4	Returning_Visitor	False	False
627.500000	0.020000	0.050000	0.0	0.0	Feb	3	3	1	4	Returning_Visitor	True	False
154.216667	0.015789	0.024561	0.0	0.0	Feb	2	2	1	3	Returning_Visitor	False	False
0.000000	0.200000	0.200000	0.0	0.4	Feb	2	4	3	3	Returning_Visitor	False	False
0.000000	0.200000	0.200000	0.0	0.0	Feb	1	2	1	5	Returning_Visitor	True	False
37.000000	0.000000	0.100000	0.0	0.8	Feb	2	2	2	3	Returning_Visitor	False	False
738.000000	0.000000	0.022222	0.0	0.4	Feb	2	4	1	2	Returning_Visitor	False	False
395.000000	0.000000	0.066667	0.0	0.0	Feb	1	1	3	3	Returning_Visitor	False	False
407.750000	0.018750	0.025833	0.0	0.4	Feb	1	1	4	3	Returning_Visitor	False	False
280.500000	0.000000	0.028571	0.0	0.0	Feb	1	1	1	3	Returning_Visitor	False	False
98.000000	0.000000	0.066667	0.0	0.0	Feb	2	5	1	3	Returning_Visitor	False	False
68.000000	0.000000	0.100000	0.0	0.0	Feb	3	2	3	3	Returning_Visitor	False	False
1668.285119	0.008333	0.016313	0.0	0.0	Feb	1	1	9	3	Returning_Visitor	False	False
0.000000	0.200000	0.200000	0.0	0.0	Feb	1	1	4	3	Returning_Visitor	False	False
334.966667	0.000000	0.007692	0.0	0.0	Feb	1	1	1	4	Returning_Visitor	True	False
32.000000	0.000000	0.100000	0.0	0.0	Feb	2	2	1	3	Returning_Visitor	False	False
2981.166667	0.000000	0.010000	0.0	0.0	Feb	2	4	4	4	Returning_Visitor	False	False

Figure 1.4: An image showing the first 20 instances of the online shoppers purchasing intention dataset

LOADING DATA

Data can be in different forms and can be available in many places. Datasets for beginners are often given in a flat format, which means that they are two-dimensional, with rows and columns. Other common forms of data may include images, **JSON** objects, and text documents. Each type of data format has to be loaded in specific ways. For example, numerical data can be loaded into memory using the **NumPy** library, which is an efficient library for working with matrices in Python.

However, we would not be able to load our marketing data **.csv** file into memory using the **NumPy** library because the dataset contains string values. For our dataset, we will use the **pandas** library because of its ability to easily work with various data types, such as strings, integers, floats, and binary values. In fact, **pandas** is dependent on **NumPy** for operations on numerical data types. **pandas** is also able to read JSON, Excel documents, and databases using SQL queries, which makes the library common among practitioners for loading and manipulating data in Python.

Here is an example of how to load a CSV file using the **NumPy** library. We use the **skiprows** argument in case there is a header, which usually contains column names:

```
import numpy as np
data = np.loadtxt(filename, delimiter=",", skiprows=1)
```

Here's an example of loading data using the **pandas** library:

```
import pandas as pd
data = pd.read_csv(filename, delimiter=",")
```

Here, we are loading in a **.CSV** file. The default delimiter is a comma, so passing this as an argument is not necessary but is useful to see. The pandas library can also handle non-numeric datatypes, which makes the library more flexible:

```
import pandas as pd
data = pd.read_json(filename)
```

The **pandas** library will flatten out the JSON and return a **DataFrame**.

The library can even connect to a database, queries can be fed directly into the function, and the table that's returned will be loaded as a **pandas** DataFrame:

```
import pandas as pd
data = pd.read_sql(con, "SELECT * FROM table")
```

We have to pass a database connection to the function in order for this to work. There is a myriad of ways for this to be achieved, depending on the database flavor.

Other forms of data that are common in deep learning, such as images and text, can also be loaded in and will be discussed later in this course.

> **NOTE**
>
> You can find all the documentation for pandas here: https://pandas.pydata.org/pandas-docs/stable/. The documentation for NumPy can be found here: https://docs.scipy.org/doc/.

EXERCISE 1.01: LOADING A DATASET FROM THE UCI MACHINE LEARNING REPOSITORY

> **NOTE**
>
> For all the exercises and activities in this chapter, you will need to have Python 3.7, Jupyter, and pandas installed on your system. Refer to the *Preface* for installation instructions. The exercises and activities are performed in Jupyter notebooks. It is recommended to keep a separate notebook for different assignments. You can download all the notebooks from this book's GitHub repository, which can be found here: https://packt.live/2OL5E9t.

In this exercise, we will be loading the **online shoppers purchasing intention** dataset from the UCI Machine Learning Repository. The goal of this exercise will be to load in the CSV data and identify a target variable to predict and the feature variables to use to model the target variable. Finally, we will separate the feature and target columns and save them to **.CSV** files so that we can use them in subsequent activities and exercises.

The dataset is related to the online behavior and activity of customers of an online store and indicates whether the user purchased any products from the website. You can find the dataset in the GitHub repository at: https://packt.live/39rdA7S.

Follow these steps to complete this exercise:

1. Open a new Jupyter Notebook and load the data into memory using the pandas library with the **read_csv** function. Import the **pandas** library and read in the **data** file:

```
import pandas as pd
data = pd.read_csv('../data/online_shoppers_intention.csv')
```

> **NOTE**
>
> The code above assumes that you are using the same folder and file structure as in the GitHub repository. If you get an error that the file cannot be found, then check to make sure your working directory is correctly structured. Alternatively, you can edit the file path in the code so that it points to the correct file location on your system, though you will need to ensure you are consistent with this when saving and loading files in later exercises.

2. To verify that we have loaded the data into the memory correctly, we can print the first few rows. Let's print out the top **20** values of the variable:

```
data.head(20)
```

The printed output should look like this:

	Administrative	Administrative_Duration	Informational	Informational_Duration	ProductRelated	ProductRelated_Duration	BounceRates	ExitRates
0	0	0.0	0	0.0	1	0.000000	0.200000	0.200000
1	0	0.0	0	0.0	2	64.000000	0.000000	0.100000
2	0	0.0	0	0.0	1	0.000000	0.200000	0.200000
3	0	0.0	0	0.0	2	2.666667	0.050000	0.140000
4	0	0.0	0	0.0	10	627.500000	0.020000	0.050000
5	0	0.0	0	0.0	19	154.216667	0.015789	0.024561
6	0	0.0	0	0.0	1	0.000000	0.200000	0.200000
7	1	0.0	0	0.0	0	0.000000	0.200000	0.200000
8	0	0.0	0	0.0	2	37.000000	0.000000	0.100000
9	0	0.0	0	0.0	3	738.000000	0.000000	0.022222
10	0	0.0	0	0.0	3	395.000000	0.000000	0.066667
11	0	0.0	0	0.0	16	407.750000	0.018750	0.025833
12	0	0.0	0	0.0	7	280.500000	0.000000	0.028571
13	0	0.0	0	0.0	6	98.000000	0.000000	0.066667
14	0	0.0	0	0.0	2	68.000000	0.000000	0.100000
15	2	53.0	0	0.0	23	1668.285119	0.008333	0.016313
16	0	0.0	0	0.0	1	0.000000	0.200000	0.200000
17	0	0.0	0	0.0	13	334.966667	0.000000	0.007692
18	0	0.0	0	0.0	2	32.000000	0.000000	0.100000
19	0	0.0	0	0.0	20	2981.166667	0.000000	0.010000

Figure 1.5: The first 20 rows and first 8 columns of the pandas DataFrame

3. We can also print the **shape** of the **DataFrame**:

```
data.shape
```

The printed output should look as follows, showing that the DataFrame has **12330** rows and **18** columns:

```
(12330, 18)
```

We have successfully loaded the data into memory, so now we can manipulate and clean the data so that a model can be trained using this data. Remember that machine learning models require data to be represented as numerical data types in order to be trained. We can see from the first few rows of the dataset that some of the columns are string types, so we will have to convert them into numerical data types later in this chapter.

4. We can see that there is a given output variable for the dataset, known as **Revenue**, which indicates whether or not the user purchased a product from the website. This seems like an appropriate target to predict, since the design of the website and the choice of the products featured may be based upon the user's behavior and whether they resulted in purchasing a particular product. Create **feature** and **target** datasets as follows, providing the **axis=1** argument:

```
feats = data.drop('Revenue', axis=1)
target = data['Revenue']
```

> **NOTE**
>
> The **axis=1** argument tells the function to drop columns rather than rows.

5. To verify that the shapes of the datasets are as expected, print out the number of **rows** and **columns** of each:

> **NOTE**
>
> The code snippet shown here uses a backslash (\ ) to split the logic across multiple lines. When the code is executed, Python will ignore the backslash, and treat the code on the next line as a direct continuation of the current line.

```
print(f'Features table has {feats.shape[0]} \
rows and {feats.shape[1]} columns')
print(f'Target table has {target.shape[0]} rows')
```

This preceding code produces the following output:

```
Features table has 12330 rows and 17 columns
Target table has 12330 rows
```

We can see two important things here that we should always verify before continuing: first, the number of rows of the **feature** DataFrame and **target** DataFrame are the same. Here, we can see that both have **12330** rows. Second, the number of columns of the feature DataFrame should be one fewer than the total DataFrame, and the target DataFrame has exactly one column.

Regarding the second point, we have to verify that the target is not contained in the feature dataset; otherwise, the model will quickly find that this is the only column needed to minimize the total error, all the way down to zero. The target column doesn't necessarily have to be one column, but for binary classification, as in our case, it will be. Remember that these machine learning models are trying to minimize some cost function in which the **target** variable will be part of that cost function, usually a difference function between the predicted value and **target** variable.

6. Finally, save the DataFrames as CSV files so that we can use them later:

```
feats.to_csv('../data/OSI_feats.csv', index=False)
target.to_csv('../data/OSI_target.csv', \
              header='Revenue', index=False)
```

> **NOTE**
>
> The **header='Revenue'** parameter is used to provide a column name. We will do this to reduce confusion later on. The **index=False** parameter is used so that the index column is not saved.

In this section, we have successfully demonstrated how to load data into Python using the **pandas** library. This will form the basis of loading data into memory for most tabular data. Images and large documents, both of which are other common forms of data for machine learning applications, have to be loaded in using other methods, all of which will be discussed later in this book.

> **NOTE**
>
> To access the source code for this specific section, please refer to https://packt.live/2YZRAyB.
>
> You can also run this example online at https://packt.live/3dVR0pF.

DATA PREPROCESSING

To fit models to the data, it must be represented in numerical format since the mathematics used in all machine learning algorithms only works on matrices of numbers (you cannot perform linear algebra on an image). This will be one goal of this section: to learn how to encode all the features into numerical representations. For example, in binary text, values that contain one of two possible values may be represented as zeros or ones. An example of this can be seen in the following diagram. Since there are only two possible values, the value **0** is assumed to be a **cat** and the value **1** is assumed to be a **dog**.

We can also rename the column for interpretation:

Label		is_dog
Cat		0
Dog		1
Cat		0
Cat		0
Dog		1
Dog		1

Figure 1.6: A numerical encoding of binary text values

Another goal will be to appropriately represent the data in numerical format - by appropriately, we mean that we want to encode relevant information numerically through the distribution of numbers. For example, one method to encode the months of the year would be to use the number of the month in the year. For example, **January** would be encoded as **1**, since it is the first month, and **December** would be **12**. Here's an example of how this would look in practice:

month		month
January		1
March		3
October		10
April		4
July		7
January		1

Figure 1.7: A numerical encoding of months

Not encoding information appropriately into numerical features can lead to machine learning models learning unintuitive representations, as well as relationships between the **feature** data and **target** variables that will prove useless for human interpretation.

An understanding of the machine learning algorithms you are looking to use will also help encode features into numerical representations appropriately. For example, algorithms for classification tasks such as Artificial Neural Networks (ANNs) and logistic regression are susceptible to large variations in the scale between the features that may hamper their model-fitting ability.

Take, for example, a regression problem attempting to fit house attributes, such as the area in square feet and the number of bedrooms, to the house price. The bounds of the area may be anywhere from **0** to **5000**, whereas the number of bedrooms may only vary from **0** to **6**, so there is a large difference between the scale of the variables.

An effective way to combat the large variation in scale between the features is to normalize the data. Normalizing the data will scale the data appropriately so that it is all of a similar magnitude. This ensures that any model coefficients or weights can be compared correctly. Algorithms such as decision trees are unaffected by data scaling, so this step can be omitted for models using tree-based algorithms.

In this section, we demonstrated a number of different ways to encode information numerically. There is a myriad of alternative techniques that can be explored elsewhere. Here, we will show some simple and popular methods that can be used to tackle common data formats.

EXERCISE 1.02: CLEANING THE DATA

It is important that we clean the data appropriately so that it can be used for training models. This often includes converting non-numerical datatypes into numerical datatypes. This will be the focus of this exercise - to convert all the columns in the feature dataset into numerical columns. To complete the exercise, perform the following steps:

1. First, we load the **feature** dataset into memory:

```
%matplotlib inline
import pandas as pd
data = pd.read_csv('../data/OSI_feats.csv')
```

2. Again, look at the first **20** rows to check out the data:

```
data.head(20)
```

The following screenshot shows the output of the preceding code:

	Administrative	Administrative_Duration	Informational	Informational_Duration	ProductRelated	ProductRelated_Duration	BounceRates	ExitRates	
0	0	0.0	0	0.0	1	0.000000	0.200000	0.200000	
1	0	0.0	0	0.0	2	64.000000	0.000000	0.100000	
2	0	0.0	0	0.0	1	0.000000	0.200000	0.200000	
3	0	0.0	0	0.0	2	2.666667	0.050000	0.140000	
4	0	0.0	0	0.0	10	627.500000	0.020000	0.050000	
5	0	0.0	0	0.0	19	154.216667	0.015789	0.024561	
6	0	0.0	0	0.0	1	0.000000	0.200000	0.200000	
7	1	0.0	0	0.0	0	0.000000	0.200000	0.200000	
8	0	0.0	0	0.0	2	37.000000	0.000000	0.100000	
9	0	0.0	0	0.0	3	738.000000	0.000000	0.022222	
10	0	0.0	0	0.0	3	395.000000	0.000000	0.066667	
11	0	0.0	0	0.0	16	407.750000	0.018750	0.025833	
12	0	0.0	0	0.0	7	280.500000	0.000000	0.028571	
13	0	0.0	0	0.0	6	98.000000	0.000000	0.066667	
14	0	0.0	0	0.0	2	68.000000	0.000000	0.100000	
15	2	53.0	0	0.0	23	1668.285119	0.008333	0.016313	
16	0	0.0	0	0.0	1	0.000000	0.200000	0.200000	
17	0	0.0	0	0.0	13	334.966667	0.000000	0.007692	
18	0	0.0	0	0.0	2	32.000000	0.000000	0.100000	
19	0	0.0	0	0.0	20	2981.166667	0.000000	0.010000	

Figure 1.8: First 20 rows and 8 columns of the pandas feature DataFrame

Here, we can see that there are a number of columns that need to be converted into the numerical format. The numerical columns we may not need to modify are the columns named **Administrative**, **Administrative_Duration**, **Informational**, **Informational_Duration**, **ProductRelated**, **ProductRelated_Duration**, **BounceRates**, **ExitRates**, **PageValues**, **SpecialDay**, **OperatingSystems**, **Browser**, **Region**, and **TrafficType**.

There is also a binary column that has either one of two possible values. This is the column named **Weekend**.

Finally, there are also categorical columns that are string types, but there are a limited number of choices (**>2**) that the column can take. These are the columns named **Month** and **VisitorType**.

3. For the numerical columns, use the **describe** function to get a quick indication of the bounds of the numerical columns:

```
data.describe()
```

The following screenshot shows the output of the preceding code:

	Administrative	Administrative_Duration	Informational	Informational_Duration	ProductRelated	ProductRelated_Duration	BounceRates	ExitRates
count	12330.000000	12330.000000	12330.000000	12330.000000	12330.000000	12330.000000	12330.000000	12330.000000
mean	2.315166	80.818611	0.503569	34.472398	31.731468	1194.746220	0.022191	0.043073
std	3.321784	176.779107	1.270156	140.749294	44.475503	1913.669288	0.048488	0.048597
min	0.000000	0.000000	0.000000	0.000000	0.000000	0.000000	0.000000	0.000000
25%	0.000000	0.000000	0.000000	0.000000	7.000000	184.137500	0.000000	0.014286
50%	1.000000	7.500000	0.000000	0.000000	18.000000	598.936905	0.003112	0.025156
75%	4.000000	93.256250	0.000000	0.000000	38.000000	1464.157214	0.016813	0.050000
max	27.000000	3398.750000	24.000000	2549.375000	705.000000	63973.522230	0.200000	0.200000

Figure 1.9: Output of the describe function in the feature DataFrame

4. Convert the binary column, **Weekend**, into a numerical column. To do so, we will examine the possible values by printing the count of each value and plotting the result, and then convert one of the values into **1** and the other into **0**. If appropriate, rename the column for interpretability.

For context, it is helpful to see the distribution of each value. We can do that using the **value_counts** function. We can try this out on the **Weekend** column:

```
data['Weekend'].value_counts()
```

We can also look at these values as a bar graph by plotting the value counts by calling the **plot** method of the resulting DataFrame and passing the **kind='bar'** argument:

```
data['Weekend'].value_counts().plot(kind='bar')
```

> **NOTE**
>
> The **kind='bar'** argument will plot the data as a bar graph. The default is a **line graph**. When plotting in Jupyter notebooks, in order to make the plots within the notebook, the following command may need to be run: **%matplotlib inline**.

The following figure shows the output of the preceding code:

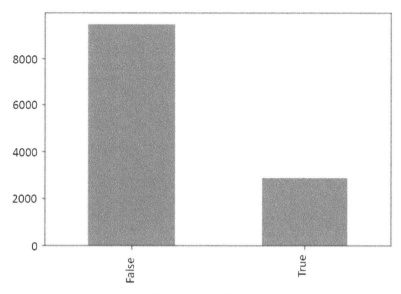

Figure 1.10: A plot of the distribution of values of the default column

5. Here, we can see that this distribution is skewed toward **false** values. This column represents whether the visit to the website occurred on a weekend, corresponding to a **true** value, or a weekday, corresponding to a **false** value. Since there are more weekdays than weekends, this skewed distribution makes sense. Convert the column into a numerical value by converting the **True** values into **1** and the **False** values into **0**. We can also change the name of the column from its default to **is_weekend**. This makes it a bit more obvious what the column means:

```
data['is_weekend'] = data['Weekend'].apply(lambda \
                     row: 1 if row == True else 0)
```

NOTE

The **apply** function iterates through each element in the column and applies the function provided as the argument. A function has to be supplied as the argument. Here, a **lambda** function is supplied.

6. Take a look at the original and converted columns side by side. Take a sample of the last few rows to see examples of both values being manipulated so that they're numerical data types:

```
data[['Weekend','is_weekend']].tail()
```

> **NOTE**
>
> The **tail** function is identical to the **head** function, except the function returns the bottom **n** values of the DataFrame instead of the top **n**.

The following figure shows the output of the preceding code:

	Weekend	is_weekend
12325	True	1
12326	True	1
12327	True	1
12328	False	0
12329	True	1

Figure 1.11: The original and manipulated column

Here, we can see that **True** is converted into **1** and **False** is converted into **0**.

7. Now we can drop the **Weekend** column, as only the **is_weekend** column is needed:

```
data.drop('Weekend', axis=1, inplace=True)
```

8. Next, we have to deal with categorical columns. We will approach the conversion of categorical columns into numerical values slightly differently than with binary text columns, but the concept will be the same. Convert each categorical column into a set of dummy columns. With dummy columns, each categorical column will be converted into **n** columns, where **n** is the number of unique values in the category. The columns will be **0** or **1**, depending on the value of the categorical column.

This is achieved with the **get_dummies** function. If we need any help understanding this function, we can use the **help** function or any function:

```
help(pd.get_dummies)
```

The following figure shows the output of the preceding code:

```
Help on function get_dummies in module pandas.core.reshape.reshape:

get_dummies(data, prefix=None, prefix_sep='_', dummy_na=False, columns=None, sparse=False, drop_first
=False, dtype=None)
    Convert categorical variable into dummy/indicator variables.

    Parameters
    ----------
    data : array-like, Series, or DataFrame
        Data of which to get dummy indicators.
    prefix : str, list of str, or dict of str, default None
        String to append DataFrame column names.
        Pass a list with length equal to the number of columns
        when calling get_dummies on a DataFrame. Alternatively, `prefix`
        can be a dictionary mapping column names to prefixes.
    prefix_sep : str, default '_'
        If appending prefix, separator/delimiter to use. Or pass a
        list or dictionary as with `prefix`.
```

Figure 1.12: The output of the help command being applied to the pd.get_dummies function

9. Let's demonstrate how to manipulate categorical columns with the **age** column. Again, it is helpful to see the distribution of values, so look at the value counts and plot them:

```
data['VisitorType'].value_counts()
data['VisitorType'].value_counts().plot(kind='bar')
```

The following figure shows the output of the preceding code:

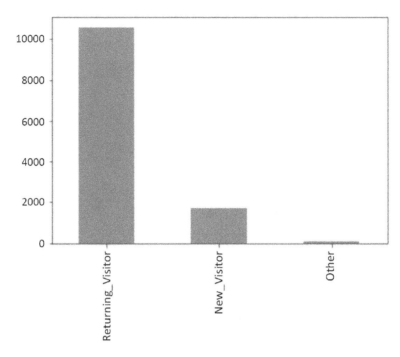

Figure 1.13: A plot of the distribution of values of the age column

10. Call the **get_dummies** function on the **VisitorType** column and take a look at the rows alongside the original:

```
colname = 'VisitorType'
visitor_type_dummies = pd.get_dummies(data[colname], \
                                      prefix=colname)
pd.concat([data[colname], \
           visitor_type_dummies], axis=1).tail(n=10)
```

The following figure shows the output of the preceding code:

	VisitorType	VisitorType_New_Visitor	VisitorType_Other	VisitorType_Returning_Visitor
12320	Returning_Visitor	0	0	1
12321	Returning_Visitor	0	0	1
12322	Returning_Visitor	0	0	1
12323	Returning_Visitor	0	0	1
12324	Returning_Visitor	0	0	1
12325	Returning_Visitor	0	0	1
12326	Returning_Visitor	0	0	1
12327	Returning_Visitor	0	0	1
12328	Returning_Visitor	0	0	1
12329	New_Visitor	1	0	0

Figure 1.14: Dummy columns from the VisitorType column

Here, we can see that, in each of the rows, there can be one value of **1**, which is in the column corresponding to the value in the **VisitorType** column.

In fact, when using dummy columns, there is some redundant information. Because we know there are three values, if two of the values in the dummy columns are **0** for a particular row, then the remaining column must be equal to **1**. It is important to eliminate any redundancy and correlations in features as it becomes difficult to determine which feature is the most important in minimizing the total error.

11. To remove the interdependency, drop the **VisitorType_Other** column because it occurs with the lowest frequency:

```
visitor_type_dummies.drop('VisitorType_Other', \
                          axis=1, inplace=True)
visitor_type_dummies.head()
```

> **NOTE**
>
> In the **drop** function, the **inplace** argument will apply the function in place, so a new variable does not have to be declared.

Looking at the first few rows, we can see what remains of our dummy columns for the original **VisitorType** column:

	VisitorType_New_Visitor	VisitorType_Returning_Visitor
0	0	1
1	0	1
2	0	1
3	0	1
4	0	1

Figure 1.15: Final dummy columns from the VisitorType column

12. Finally, add these dummy columns to the original feature data by concatenating the two DataFrames column-wise and dropping the original column:

```
data = pd.concat([data, visitor_type_dummies], axis=1)
data.drop('VisitorType', axis=1, inplace=True)
```

13. Repeat the exact same steps with the remaining categorical column, **Month**. First, examine the distribution of column values, which is an optional step. Second, create dummy columns. Third, drop one of the columns to remove redundancy. Fourth, concatenate the dummy columns into a feature dataset. Finally, drop the original column if it remains in the dataset. You can do this using the following code:

```
colname = 'Month'
month_dummies = pd.get_dummies(data[colname], prefix=colname)
month_dummies.drop(colname+'_Feb', axis=1, inplace=True)
data = pd.concat([data, month_dummies], axis=1)
data.drop('Month', axis=1, inplace=True)
```

14. Now, we should have our entire dataset as numerical columns. Check the types of each column to verify this:

```
data.dtypes
```

The following figure shows the output of the preceding code:

```
Administrative                      int64
Administrative_Duration           float64
Informational                       int64
Informational_Duration            float64
ProductRelated                      int64
ProductRelated_Duration           float64
BounceRates                       float64
ExitRates                         float64
PageValues                        float64
SpecialDay                        float64
OperatingSystems                    int64
Browser                             int64
Region                              int64
TrafficType                         int64
is_weekend                          int64
VisitorType_New_Visitor             uint8
VisitorType_Returning_Visitor       uint8
Month_Aug                           uint8
Month_Dec                           uint8
Month_Jul                           uint8
Month_June                          uint8
Month_Mar                           uint8
Month_May                           uint8
Month_Nov                           uint8
Month_Oct                           uint8
Month_Sep                           uint8
dtype: object
```

Figure 1.16: The datatypes of the processed feature dataset

15. Now that we have verified the datatypes, we have a dataset we can use to train a model, so let's save this for later:

```
data.to_csv('../data/OSI_feats_e2.csv', index=False)
```

16. Let's do the same for the **target** variable. First, load the data in, convert the column into a numerical datatype, and save the column as a CSV file:

```
target = pd.read_csv('../data/OSI_target.csv')
target.head(n=10)
```

The following figure shows the output of the preceding code:

	Revenue
0	False
1	False
2	False
3	False
4	False
5	False
6	False
7	False
8	False
9	False

Figure 1.17: First 10 rows of the target dataset

Here, we can see that this is a **Boolean** datatype and that there are two unique values.

17. Convert this into a binary numerical column, much like we did with the binary columns in the feature dataset:

```
target['Revenue'] = target['Revenue'].apply(lambda row: 1 \
                        if row==True else 0)
target.head(n=10)
```

The following figure shows the output of the preceding code:

	Revenue
0	0
1	0
2	0
3	0
4	0
5	0
6	0
7	0
8	0
9	0

Figure 1.18: First 10 rows of the target dataset when converted into integers

18. Finally, save the target dataset to a CSV file:

```
target.to_csv('../data/OSI_target_e2.csv', index=False)
```

In this exercise, we learned how to clean the data appropriately so that it can be used to train models. We converted the non-numerical datatypes into numerical datatypes; that is, we converted all the columns in the feature dataset into numerical columns. Lastly, we saved the target dataset as a CSV file so that we can use it in the following exercises and activities.

NOTE

To access the source code for this specific section, please refer to https://packt.live/2YW1DVi.

You can also run this example online at https://packt.live/2BpO4EI.

APPROPRIATE REPRESENTATION OF THE DATA

In our online shoppers purchase intention dataset, we have some columns that are defined as numerical variables when, upon closer inspection, they are actually categorical variables that have been given numerical labels. These columns are **OperatingSystems**, **Browser**, **TrafficType**, and **Region**. Currently, we have treated them as numerical variables, though they are categorical, which should be encoded into the features if we want the models we build to learn the relationships between the features and the target.

We do this because we may be encoding some misleading relationships in the features. For example, if the value of the **OperatingSystems** field is equal to **2**, does that mean it is twice the value as that which has the value **1**? Probably not, since it refers to the operating system. For this reason, we will convert the field into a categorical variable. The same may be applied to the **Browser**, **TrafficType**, and **Region** columns.

EXERCISE 1.03: APPROPRIATE REPRESENTATION OF THE DATA

In this exercise, we will convert the **OperatingSystems**, **Browser**, **TrafficType**, and **Region** columns into categorical types to accurately reflect the information. To do this, we will create dummy variables from the column in a similar manner to what we did in *Exercise 1.02*, *Cleaning the Data*. To do so, perform the following steps:

1. Open a Jupyter Notebook.

2. Load the dataset into memory. We can use the same feature dataset that was the output from *Exercise 1.02*, *Cleaning the Data*, which contains the original numerical versions of the **OperatingSystems**, **Browser**, **TrafficType**, and **Region** columns:

```
import pandas as pd
data = pd.read_csv('../data/OSI_feats_e2.csv')
```

3. Look at the distribution of values in the **OperatingSystems** column:

```
data['OperatingSystems'].value_counts()
```

The following figure shows the output of the preceding code:

```
2    6601
1    2585
3    2555
4     478
8      79
6      19
7       7
5       6
Name: OperatingSystems, dtype: int64
```

Figure 1.19: The distribution of values in the OperatingSystems column

4. Create dummy variables from the **OperatingSystem** column:

```
colname = 'OperatingSystems'
operation_system_dummies = pd.get_dummies(data[colname], \
                    prefix=colname)
```

5. Drop the dummy variable representing the value with the lowest occurring frequency and join back with the original data:

```
operation_system_dummies.drop(colname+'_5', axis=1, \
                    inplace=True)
data = pd.concat([data, operation_system_dummies], axis=1)
```

6. Repeat this for the **Browser** column:

```
data['Browser'].value_counts()
```

The following figure shows the output of the preceding code:

```
2      7961
1      2462
4       736
5       467
6       174
10      163
8       135
3       105
13       61
7        49
12       10
11        6
9         1
Name: Browser, dtype: int64
```

Figure 1.20: The distribution of values in the Browser column

7. Create dummy variables, drop the dummy variable with the lowest occurring frequency, and join back with the original data:

```
colname = 'Browser'
browser_dummies = pd.get_dummies(data[colname], \
            prefix=colname)
browser_dummies.drop(colname+'_9', axis=1, inplace=True)
data = pd.concat([data, browser_dummies], axis=1)
```

8. Repeat this for the **TrafficType** and **Region** columns:

> **NOTE**
>
> The # symbol in the code snippet below denotes a code comment. Comments are added into code to help explain specific bits of logic.

```
colname = 'TrafficType'
data[colname].value_counts()
traffic_dummies = pd.get_dummies(data[colname], prefix=colname)
# value 17 occurs with lowest frequency
```

```
traffic_dummies.drop(colname+'_17', axis=1, inplace=True)
data = pd.concat([data, traffic_dummies], axis=1)


colname = 'Region'
data[colname].value_counts()
region_dummies = pd.get_dummies(data[colname], \
                    prefix=colname)
# value 5 occurs with lowest frequency
region_dummies.drop(colname+'_5', axis=1, inplace=True)
data = pd.concat([data, region_dummies], axis=1)
```

9. Check the column types to verify they are all numerical:

```
data.dtypes
```

The following figure shows the output of the preceding code:

```
Administrative               int64
Administrative_Duration      float64
Informational                int64
Informational_Duration       float64
ProductRelated               int64
                             ...
Region_4                     uint8
Region_6                     uint8
Region_7                     uint8
Region_8                     uint8
Region_9                     uint8
Length: 68, dtype: object
```

Figure 1.21: The datatypes of the processed feature dataset

10. Finally, save the dataset to a CSV file for later use:

```
data.to_csv('../data/OSI_feats_e3.csv', index=False)
```

Now, we can accurately test whether the browser type, operating system, traffic type, or region will affect the target variable. This exercise has demonstrated how to appropriately represent data for use in machine learning algorithms. We have presented some techniques that we can use to convert data into numerical datatypes that cover many situations that may be encountered when working with tabular data.

> **NOTE**
>
> To access the source code for this specific section, please refer to https://packt.live/3dXOTBy.
>
> You can also run this example online at https://packt.live/3iBvDxw.

LIFE CYCLE OF MODEL CREATION

In this section, we will cover the life cycle of creating performant machine learning models, from engineering features to fitting models to training data, and evaluating our models using various metrics. The following diagram demonstrates the iterative process of building machine learning models. Features are engineered that represent potential correlations between the features and the target, the model is fit, and then models are evaluated.

Depending on how the model is scored according to the model's evaluation metrics, the features are engineered further, and the process continues. Many of the steps that are implemented to create models are highly transferable between all machine learning libraries. We'll start with scikit-learn, which has the advantage of being widely used, and as such, there is a lot of documentation, tutorials, and learning materials to be found across the internet:

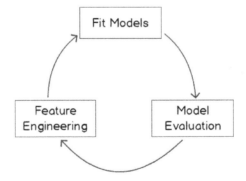

Figure 1.22: The life cycle of model creation

MACHINE LEARNING LIBRARIES

While this book is an introduction to deep learning with Keras, as we mentioned earlier, we will start by utilizing scikit-learn. This will help us establish the fundamentals of building a machine learning model using the Python programming language.

Similar to scikit-learn, Keras makes it easy to create models in the Python programming language through an easy-to-use API. However, the goal of Keras is the creation and training of neural networks, rather than machine learning models in general. ANNs represent a large class of machine learning algorithms, and they are so-called because their architecture resembles the neurons in the human brain. The Keras library has many general-purpose functions built-in, such as **optimizers**, **activation functions**, and **layer properties**, so that users, like in scikit-learn, do not have to code these algorithms from scratch.

SCIKIT-LEARN

Scikit-learn was initially created by David Cournapeau in 2007 as a way to easily create machine learning models in the Python programming language. Since its inception, the library has grown immensely in popularity because of its ease of use, wide adoption within the machine learning community, and flexibility of use. scikit-learn is usually the first machine learning package that's implemented by practitioners using Python because of the large number of algorithms available for **classification**, **regression**, and **clustering** tasks and the speed with which results can be obtained.

For example, scikit-learn's **LinearRegression** class is an excellent choice if you wish to quickly train a simple regression model, whereas if a more complex algorithm is required that's capable of learning nonlinear relationships, scikit-learn's **GradientBoostingRegressor** or any one of the **support vector machine** algorithms are great choices. Likewise, with classification or clustering tasks, scikit-learn offers a wide variety of algorithms to choose from.

The following are a few of the advantages and disadvantages of using scikit-learn for machine learning purposes.

The advantages of scikit-learn are as follows:

- **Mature**: Scikit-learn is well-established within the community and used by members of the community of all skill levels. The package includes most of the common machine learning algorithms for classification, regression, and clustering tasks.

- **User-friendly**: Scikit-learn features an easy-to-use API that allows beginners to efficiently prototype without having to have a deep understanding or having to code each specific mode.

- **Open source**: There is an active open source community working to improve the library, add documentation, and release regular updates, which ensures that the package is stable and up to date.

The disadvantage of scikit-learn is as follows:

Neural network support is lacking: Estimators with ANN algorithms are minimal.

> **NOTE**
>
> You can find all the documentation for the scikit-learn library here: https://scikit-learn.org/stable/documentation.html.

The estimators in scikit-learn can generally be classified into supervised learning and unsupervised learning techniques. Supervised learning occurs when a `target` variable is present. A `target` variable is a variable of the dataset that you are trying to predict, given the other variables. `Supervised learning` requires the target variable to be known and models are trained to correctly predict this variable. `Binary classification` using `logistic regression` is a good example of a supervised learning technique.

In `unsupervised learning`, no target variable is given in the training data, but models aim to assign a target variable. An example of an unsupervised learning technique is k-means clustering. This algorithm partitions data into a given number of clusters based on its proximity to neighboring data points. The `target` variable that's assigned may be either the cluster number or cluster center.

An example of utilizing a clustering example in practice may look as follows. Imagine that you are a jacket manufacturer and your goal is to develop dimensions for various jacket sizes. You cannot create a custom-fit jacket for each customer, so one option you have to determine the dimensions for jackets is to sample the population of customers for various parameters that may be correlated to fit, such as height and weight. Then, you can group the population into clusters using scikit-learn's `k-means clustering` algorithm with a cluster number that matches the number of jacket sizes you wish to produce. The cluster-centers that are created from the clustering algorithm become the parameters that the jacket sizes are based on.

This is visualized in the following figure:

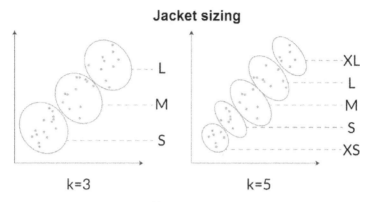

Figure 1.23: An unsupervised learning example of grouping customer
parameters into clusters

There are even **semi-supervised learning** techniques in which unlabeled data is used in the training of machine learning models. This technique may be used if there is only a small amount of labeled data and a copious amount of unlabeled data. In practice, semi-supervised learning produces a significant improvement in model performance compared to unsupervised learning.

The scikit-learn library is ideal for beginners as the general concepts for building machine learning pipelines can be learned easily. Concepts such as data preprocessing (the preparation of data for use in machine learning models), hyperparameter tuning (the process of selecting the appropriate model parameters), model evaluation (the quantitative evaluation of a model's performance), and many more are all included in the library. Even experienced users find the library easy to use in order to rapidly prototype models before using a more specialized machine learning library.

Indeed, the various machine learning techniques we've discussed, such as supervised and unsupervised learning, can be applied with Keras using neural networks with different architectures, all of which will be discussed throughout this book.

KERAS

Keras is designed to be a high-level neural network API that is built on top of frameworks such as TensorFlow, CNTK, and Theano. One of the great benefits of using Keras as an introduction to deep learning for beginners is that it is very user-friendly; advanced functions such as optimizers and layers are already built into the library and do not have to be written from scratch. This is why Keras is popular not only among beginners but also seasoned experts. Also, the library allows the rapid prototyping of neural networks, supports a wide variety of network architectures, and can be run on both CPUs and GPUs.

> **NOTE**
>
> You can find the library and all the documentation for Keras here: https://Keras.io/.

Keras is used to create and train neural networks and does not offer much in terms of other machine learning algorithms, including supervised algorithms such as support vector machines and unsupervised algorithms such as `k-means clustering`. What Keras does offer, though, is a well-designed API that can be used to create and train neural networks, which takes away much of the effort that's required to apply linear algebra and multivariate calculus accurately.

The specific modules that are available from the Keras library, such as `neural layers`, `cost functions`, `optimizers`, `initialization schemes`, `activation functions`, and `regularization schemes`, will be explained thoroughly throughout this book. All these modules have relevant functions that can be used to optimize performance for training neural networks for specific tasks.

ADVANTAGES OF KERAS

Here are a few of the main advantages of using Keras for machine learning purposes:

- **User-friendly**: Much like scikit-learn, Keras features an easy-to-use API that allows users to focus on model-building rather than the specifics of the algorithms.

- **Modular**: The API consists of fully configurable modules that can all be plugged together and work seamlessly.

- **Extensible**: It is relatively simple to add new modules to the library. This allows users to take advantage of the many robust modules within the library while providing them the flexibility to create their own.

- **Open source**: Keras is an open source library and is constantly improving and adding modules to its code base thanks to the work of many collaborators working in conjunction to build improvements and help create a robust library for all.

- **Works with Python**: Keras models are declared directly in Python rather than in separate configuration files, which allows Keras to take advantage of working with Python, such as ease of debugging and extensibility.

DISADVANTAGES OF KERAS

Here are a few of the main disadvantages of using Keras for machine learning purposes:

- **Advanced customization**: While simple surface-level customization such as creating simple custom loss functions or neural layers is facile, it can be difficult to change how the underlying architecture works.

- **Lack of examples**: Beginners often rely on examples to kick-start their learning. Advanced examples can be lacking in the Keras documentation, which can prevent beginners from advancing in their learning.

Keras offers those familiar with the Python programming language and machine learning the ability to create neural network architectures easily. Since neural networks are quite complicated, we will use scikit-learn to introduce many machine learning concepts before applying them to the Keras library.

MORE THAN BUILDING MODELS

While machine learning libraries such as scikit-learn and Keras were created to help build and train predictive models, their practicality extends much further. One common use case of building models is that they can be utilized to perform predictions on new data. Once a model has been trained, new observations can be fed into the model to generate predictions. Models may even be used as intermediate steps. For example, neural network models can be used as **feature extractors**, classifying objects in an image that can then be fed into a subsequent model, as illustrated in the following image:

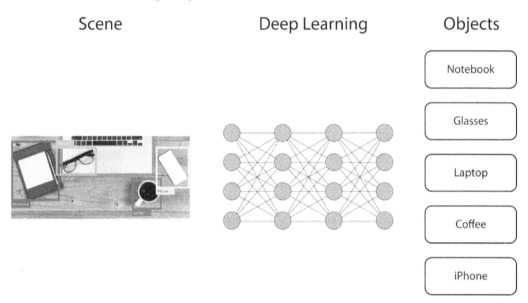

Figure 1.24: Classifying objects using deep learning

Another common use case for models is that they can be used to summarize datasets by learning representations of the data. Such models are known as auto-encoders, a type of neural network architecture that can be used to learn such representations of a given dataset. Therefore, the dataset can thus be represented in a reduced dimension with minimal loss of information:

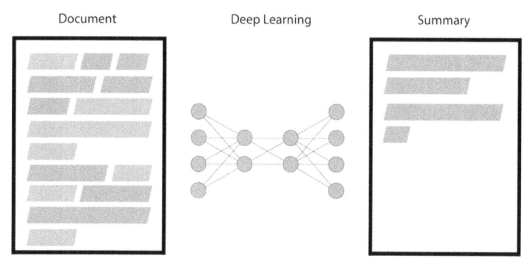

Figure 1.25: An example of using deep learning for text summarization

MODEL TRAINING

In this section, we will begin fitting our model to the datasets that we have created. In this chapter, we will review the minimum steps that are required to create a machine learning model that can be applied when building models with any machine learning library, including scikit-learn and Keras.

CLASSIFIERS AND REGRESSION MODELS

This book is concerned with applications of deep learning. The vast majority of deep learning tasks are supervised learning, in which there is a given target, and we want to fit a model so that we can understand the relationship between the features and the target.

An example of supervised learning is identifying whether a picture contains a **dog** or a **cat**. We want to determine the relationship between the **input** (a matrix of pixel values) and the **target** variable, that is, whether the image is of a **dog** or a **cat**:

Figure 1.26: A simple supervised learning task to classify images as dogs and cats

Of course, we may need many more images in our training dataset to robustly classify new images, but models that are trained on such a dataset are able to identify the various relationships that differentiate cats and dogs, which can then be used to predict labels for new data.

Supervised learning models are generally used for either classification or regression tasks.

CLASSIFICATION TASKS

The goal of classification tasks is to fit models from data with discrete categories that can be used to label `unlabeled data`. For example, these types of models can be used to classify images as dogs or cats. But it doesn't stop at binary classification; multi-label classification is also possible. Another example of how this may be a `classification` task would be to predict the existence of dogs within the images. A positive prediction would indicate the presence of dogs within the images, while a negative prediction would indicate no presence of dogs. Note that this could easily be converted into a `regression` task, that is, the estimation of a continuous variable as opposed to a discrete variable, which classification tasks estimate, by predicting the number of dogs within the images.

Most classification tasks output a probability for each unique class. This prediction is determined as the class with the highest probability, as can be seen in the following figure:

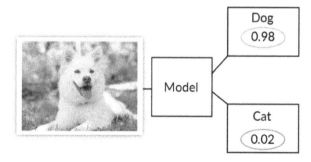

Figure 1.27: An illustration of a classification model labeling an image

Some of the most common classification algorithms are as follows:

- **Logistic regression**: This algorithm is similar to linear regression, in which feature coefficients are learned and predictions are made by taking the sum of the product of the feature coefficients and features.

- **Decision trees**: This algorithm follows a tree-like structure. Decisions are made at each node and branches represent possible options at the node, terminating in the predicted result.

- **ANNs**: ANNs replicate the structure and performance of a biological neural network to perform pattern recognition tasks. An ANN consists of interconnected neurons, laid out with a set architecture, that pass information to each other until a result is achieved.

REGRESSION TASKS

While the aim of `classification` tasks is to label datasets with discrete variables, the aim of `regression` tasks is to provide input data with continuous variables and output a numerical value. For example, if you have a dataset of stock market prices, a classification task may predict whether to buy, sell, or hold, whereas a regression task will predict what the stock market price will be.

A simple yet very popular algorithm for regression tasks is linear regression. It consists of only one independent feature (**x**), whose relationship with its dependent feature (**y**) is linear. Due to its simplicity, it is often overlooked, even though it performs very well for simple data problems.

Some of the most common regression algorithms are as follows:

- **Linear regression**: This algorithm learns feature coefficients and predictions are made by taking the sum of the product of the feature coefficients and features.

- **Support Vector Machines**: This algorithm uses kernels to map input data into a multi-dimensional feature space to understand relationships between features and the target.

- **ANNs**: ANNs replicate the structure and performance of a biological neural network to perform pattern recognition tasks. An ANN consists of interconnected neurons, laid out with a set architecture, that pass information to each other until a result is achieved.

TRAINING DATASETS AND TEST DATASETS

Whenever we create machine learning models, we separate the data into **training** and **test** datasets. The training data is the set of data that's used to train the model. Typically, it is a large proportion—around **80%**—of the total dataset. The test dataset is a sample of the dataset that is held out from the beginning and is used to provide an unbiased evaluation of the model. The test dataset should represent real-world data as accurately as possible. Any model evaluation metrics that are reported should be applied to the test dataset unless it's explicitly stated that the metrics have been evaluated on the training dataset. The reason for this is that models will typically perform better on the data they are trained on.

Furthermore, models can overfit the training dataset, meaning that they perform well on the training dataset but perform poorly on the **test** dataset. A model is said to be overfitted to the data if the model's performance is very good when evaluated on the **training** dataset, but it performs poorly on the **test** dataset. Conversely, a model can be underfitted to the data. In this case, the model will fail to learn relationships between the **features** and the **target**, which will lead to poor performance when evaluated on both the **training** and **test** datasets.

We aim for a balance of the two, not relying so heavily on the **training** dataset that we overfit but allowing the model to learn the relationships between the **features** and the **target** so that the model generalizes well to new data. This concept is illustrated in the following figure:

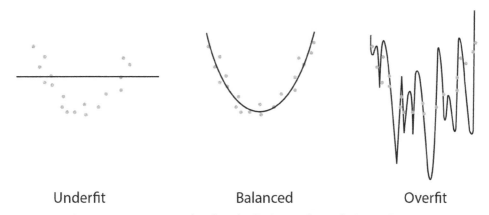

Underfit   Balanced   Overfit

Figure 1.28: An example of underfitting and overfitting a dataset

There are many ways to split the dataset via **sampling** methods. One way to split a dataset into training is to simply randomly sample the data until you have the desired number of data points. This is often the default method in functions such as the scikit-learn **train_test_spilt** function.

Another method is to stratify the sampling. In stratified sampling, each subpopulation is sampled independently. Each subpopulation is determined by the target variable. This can be advantageous in examples such as **binary classification**, where the target variable is highly skewed toward one value or another, and random sampling may not provide data points of both values in the **training** and **test** datasets. There are also validation datasets, which we will address later in this chapter.

MODEL EVALUATION METRICS

It is important to be able to evaluate our models effectively, not just in terms of the model's performance but also in the context of the problem we are trying to solve. For example, let's say we built a **classification** task to predict whether to buy, sell, or hold stock based on historical stock market prices. If our model only predicted to buy every time, this would not be a useful result because we may not have infinite resources to buy stock. It may be better to be less accurate yet also include some sell predictions.

Common evaluation metrics for **classification** tasks include accuracy, precision, recall, and f1 score. **Accuracy** is defined as the number of correct predictions divided by the total number of predictions. **Accuracy** is very interpretable and relatable and good for when there are balanced classes. When the classes are highly skewed, accuracy can be misleading, however:

$$\text{Accuracy} = \frac{\text{Number of correct predictions}}{\text{Total number of predictions}}$$

Figure 1.29: Formula to calculate accuracy

Precision is another useful metric. It's defined as the number of true positive results divided by the total number of positive results (true and false) predicted by the model:

$$\text{Precision} = \frac{\text{True Positives}}{\text{True Positives} + \text{False Positives}}$$

Figure 1.30: Formula to calculate precision

Recall is defined as the number of correct positive results divided by all the positive results from the ground truth:

$$\text{Recall} = \frac{\text{True Positives}}{\text{True Positives} + \text{False Negatives}}$$

Figure 1.31: Formula to calculate recall

Both **precision** and **recall** are scored between **zero** and **one** but scoring well on one may mean scoring poorly on the other. For example, a model may have high precision but low recall, which indicates that the model is very accurate but misses a large number of positive instances. It is useful to have a metric that combines recall and precision. Enter the **F1 score**, which determines how precise and robust your model is:

$$\text{F1 Score} = 2 \times \frac{1}{\frac{1}{\text{Precision}} + \frac{1}{\text{Recall}}}$$

Figure 1.32: Formula to calculate the F1 score

When evaluating models, it is helpful to look at a range of different evaluation metrics. They will help determine the most appropriate model and evaluate where the model is misclassifying predictions.

For example, take a model that helps doctors predict the presence of a rare disease in their patients. By predicting a negative result for every instance, the model might provide a highly accurate evaluation, but this would not help the doctors or patients very much. Instead, examining the **precision** or **recall** may be much more informative.

A high precision model is very picky and will likely ensure that all predictions labeled positive are indeed positive. A high recall model is likely to recall many of the **true** positive instances, at the cost of incurring many false positives.

A **high precision model** is desired when you want to be sure the predictions labeled as **true** have a high likelihood of being true. In our example, this may be desired if the cost of treating a rare disease or risk of treatment complications is high. A **high recall model** is desired if you want to make sure your model recalls as many **true** positives as possible. In our example, this may be the case if the rare disease is highly contagious and we want to be sure all cases of the disease are treated.

EXERCISE 1.04: CREATING A SIMPLE MODEL

In this exercise, we will create a simple **logistic regression model** from the **scikit-learn** package. Then, we will create some model evaluation metrics and test the predictions against those model evaluation metrics.

We should always approach training any machine learning model as an iterative approach, beginning with a simple model and using model evaluation metrics to evaluate the performance of the models. In this model, our goal is to classify the users in the online shoppers purchasing intention dataset into those that will purchase during their session and those that will not. Follow these steps to complete this exercise:

1. Load in the data:

```
import pandas as pd
feats = pd.read_csv('../data/OSI_feats_e3.csv')
target = pd.read_csv('../data/OSI_target_e2.csv')
```

2. Begin by creating a **test** and **training** dataset. Train the data using the **training** dataset and evaluate the performance of the model on the **test** dataset.

    We will use **test_size = 0.2**, which means that **20%** of the data will be reserved for testing, and we will set a number for the **random_ state** parameter:

```
from sklearn.model_selection import train_test_split
test_size = 0.2
random_state = 42
X_train, X_test, \
y_train, y_test = train_test_split(feats, target, \
                                   test_size=test_size, \
                                   random_state=random_state)
```

3. Print out the **shape** of each DataFrame to verify that the dimensions are correct:

```
print(f'Shape of X_train: {X_train.shape}')
print(f'Shape of y_train: {y_train.shape}')
print(f'Shape of X_test: {X_test.shape}')
print(f'Shape of y_test: {y_test.shape}')
```

The preceding code produces the following output:

```
Shape of X_train: (9864, 68)
Shape of y_train: (9864, 1)
Shape of X_test: (2466, 68)
Shape of y_test: (2466, 1)
```

These dimensions look correct; each of the **target** datasets has a single column, the training feature and **target** DataFrames have the same number of rows, the same applies to the **test** feature and **target** DataFrames, and the test DataFrames are **20%** of the total dataset.

4. Next, instantiate the model:

```
from sklearn.linear_model import LogisticRegression
model = LogisticRegression(random_state=42)
```

While there are many arguments we can add to scikit-learn's logistic regression model (such as the type and value of the regularization parameter, the type of solver, and the maximum number of iterations for the model to have), we will only pass **random_state**.

5. Then, **fit** the model to the training data:

```
model.fit(X_train, y_train['Revenue'])
```

6. To test the performance of the model, compare the predictions of the model with the true values:

```
y_pred = model.predict(X_test)
```

There are many types of model evaluation metrics that we can use. Let's start with the **accuracy**, which is defined as the proportion of predicted values that equal the true values:

```
from sklearn import metrics
accuracy = metrics.accuracy_score(y_pred=y_pred, \
                                  y_true=y_test)
print(f'Accuracy of the model is {accuracy*100:.4f}%')
```

The preceding code produces the following output:

```
Accuracy of the model is 87.0641%
```

7. Other common evaluation metrics for classification models include **precision**, **recall**, and **fscore**. Use the scikit-learn **precison_recall_fscore_support** function, which can calculate all three:

```
precision, recall, fscore, _ = \
metrics.precision_recall_fscore_support(y_pred=y_pred, \
                                        y_true=y_test, \
                                        average='binary')
print(f'Precision: {precision:.4f}\nRecall: \
{recall:.4f}\nfscore: {fscore:.4f}')
```

> **NOTE**
>
> The underscore is used in Python for many reasons. It can be used to recall the value of the last expression in the interpreter, but in this case, we're using it to ignore specific values that are output by the function.

The following figure shows the output of the preceding code:

```
Precision: 0.7347
Recall: 0.3504
fscore: 0.4745
```

Since these metrics are scored between **0** and **1**, the **recall** and **fscore** are not as impressive as the **accuracy**, though looking at all of these metrics together can help us find where our models are doing well and where they could be improved by examining in which observations the model gets predictions incorrect.

8. Look at the coefficients that the model outputs to observe which features have a greater impact on the overall result of the prediction:

```
coef_list = [f'{feature}: {coef}' for coef, \
             feature in sorted(zip(model.coef_[0], \
             X_train.columns.values.tolist()))]
for item in coef_list:
    print(item)
```

The following figure shows the output of the preceding code:

```
TrafficType_13: -0.9393317018656502
VisitorType_Returning_Visitor: -0.7126379729869377
Month_Dec: -0.6356666079086347
ExitRates: -0.6168306621684505
Month_Mar: -0.5531772345591857
Region_9: -0.5493990371550316
TrafficType_3: -0.5230504004211978
OperatingSystems_3: -0.5047311736766499
SpecialDay: -0.48888883272346506
BounceRates: -0.4573686067908481
Month_May: -0.4436363104925222
Month_June: -0.4225194836012355
OperatingSystems_8: -0.35057329371369783
Browser_6: -0.33033671140440707
TrafficType_6: -0.2572321108188088
TrafficType_1: -0.24969535181259417
Browser_3: -0.23765128996809284
VisitorType_New_Visitor: -0.22945892368475135
Browser_1: -0.22069737949723414
Region_7: -0.21116529737609177
Browser_13: -0.20773332314846657
Region_4: -0.20645936733062473
Browser_4: -0.18452552602906916
OperatingSystems_4: -0.17537032410289136
OperatingSystems_2: -0.17087815382440244
OperatingSystems_1: -0.14530926674716454
TrafficType_15: -0.12601954689866632
TrafficType_4: -0.12551302296797587
Browser_2: -0.12254444691952127
Region_3: -0.116409339032699
TrafficType_9: -0.09345050196986791
Browser_8: -0.07432180699436479
Browser_5: -0.06731941488695285
TrafficType_19: -0.04763319631540111
Browser_10: -0.03030326779492614
TrafficType_14: -0.02486754694456821
Region_1: -0.024392989712640506
TrafficType_18: -0.02222257922449895
TrafficType_20: -0.01833180070703584155
OperatingSystems_6: -0.016786449649954342
TrafficType_7: -0.006542353054798274
TrafficType_12: -0.0032342542351401346
Browser_11: -0.002452753984304908
Informational_Duration: -0.00032045144921367014
Administrative_Duration: -0.00010008862449623993
ProductRelated_Duration: 4.6077899325827885e-05
ProductRelated: 0.003291131517956643
Administrative: 0.008809132521965357
TrafficType_2: 0.025894902253396974
Browser_7: 0.028686788285342275
Region_8: 0.029319493036519817
OperatingSystems_7: 0.03298640042309421
TrafficType_16: 0.047341484936212506
Informational: 0.08555002045301442
TrafficType_5: 0.0859889420171317
PageValues: 0.08672528112710322
Region_6: 0.09309020409318655
Month_Aug: 0.09668425308005028
Browser_12: 0.1189651797379178
is_weekend: 0.11966844048422016
Month_Sep: 0.12544889935651957
Region_2: 0.13313545468089413
TrafficType_11: 0.19223716898106263
Month_Jul: 0.21082793061040983
Month_Oct: 0.2715030204884287
TrafficType_10: 0.35298265536282414
TrafficType_8: 0.4020350043660541
Month_Nov: 0.5044070793869467
```

Figure 1.33: The sorted important features of the model with their respective coefficients

This exercise has taught us how to create and train a predictive model to predict a **target** variable when given **feature** variables. We split the **feature** and **target** dataset into **training** and **test** datasets. Then, we trained our model on the **training** dataset and evaluated our model on the **test** dataset. Finally, we observed the trained coefficients for this model.

> **NOTE**
>
> To access the source code for this specific section, please refer to https://packt.live/2Aq3ZCc.
>
> You can also run this example online at https://packt.live/2VIRSaL.

MODEL TUNING

In this section, we will delve further into evaluating model performance and examine techniques that we can use to generalize models to new data using **regularization**. Providing the context of a model's performance is extremely important. Our aim is to determine whether our model is performing well compared to trivial or obvious approaches. We do this by creating a baseline model against which machine learning models we train are compared. It is important to stress that all model evaluation metrics are evaluated and reported via the **test** dataset since that will give us an understanding of how the model will perform on new data.

BASELINE MODELS

A baseline model should be a simple and well-understood procedure, and the performance of this model should be the lowest acceptable performance for any model we build. For classification models, a useful and easy baseline model is to calculate the model outcome value. For example, if there are **60% false** values, our baseline model would be to predict false for every value, which would give us an **accuracy** of **60%**. For **regression models**, the **mean** or **median** can be used as the baseline.

EXERCISE 1.05: DETERMINING A BASELINE MODEL

In this exercise, we will put the model performance into context. The accuracy we attained from our model seemed good, but we need something to compare it to. Since machine learning model performance is relative, it is important to develop a robust baseline with which to compare models. Once again, we are using the online shoppers purchasing intention dataset, and our **target** variable is whether or not each user will purchase a product in their session. Follow these steps to complete this exercise:

1. Import the **pandas** library and load in the **target** dataset:

```
import pandas as pd
target = pd.read_csv('../data/OSI_target_e2.csv')
```

2. Next, calculate the relative proportion of each value of the **target** variables:

```
target['Revenue'].value_counts()/target.shape[0]*100
```

The following figure shows the output of the preceding code:

```
0       84.525547
1       15.474453
Name: Revenue, dtype: float64
```

Figure 1.34: Relative proportion of each value

3. Here, we can see that **0** is represented **84.525547%** of the time—that is, there is no purchase by the user, and this is our **baseline** accuracy. Now, for the other model evaluation metrics:

```
from sklearn import metrics
y_baseline = pd.Series(data=[0]*target.shape[0])
precision, recall, \
fscore, _ = metrics.precision_recall_fscore_support\
            (y_pred=y_baseline, \
             y_true=target['Revenue'], average='macro')
```

Here, we've set the baseline model to predict **0** and have repeated the value so that it's the same as the number of rows in the **test** dataset.

> **NOTE**
>
> The average parameter in the **precision_recall_fscore_support** function has to be set to **macro** because when it is set to **binary**, as it was previously, the function is looking for **true** values, and our **baseline** model only consists of **false** values.

4. Print the final output for precision, recall, and fscore:

```
print(f'Precision: {precision:.4f}\nRecall:\
{recall:.4f}\nfscore: {fscore:.4f}')
```

The preceding code produces the following output:

```
Precision: 0.9226
Recall: 0.5000
Fscore: 0.4581
```

Now, we have a baseline model that we can compare to our previous model, as well as any subsequent models. By doing this, we can tell that while the accuracy of our previous model seemed high, it did not score much better than this **baseline** model.

> **NOTE**
>
> To access the source code for this specific section, please refer to https://packt.live/31MD1jH.
>
> You can also run this example online at https://packt.live/2VFFSXO.

REGULARIZATION

Earlier in this chapter, we learned about **overfitting** and what it looks like. The hallmark of **overfitting** is when a model is trained on the training data and performs extremely well yet performs terribly on **test** data. One reason for this could be that the model may be relying too heavily on certain features that lead to good performance in the training dataset but do not generalize well to new observations of data or the test dataset.

One technique that can be used to avoid this is called **regularization**. Regularization constrains the values of the coefficients toward zero, which discourages a complex model. There are many different types of regularization techniques. For example, in **linear** and **logistic** regression, **ridge** and **lasso** regularization are most common. In tree-based models, limiting the maximum depth of the trees acts as regularization.

There are two different types of regularization, namely **L1** and **L2**. This term is either the **L2** norm (the sum of the squared values) of the weights or the **L1** norm (the sum of the absolute values) of the weights. Since the **l1** regularization parameter acts as a feature selector, it is able to reduce the coefficient of features to zero. We can use the output of this model to observe which features do not contribute much to the performance and remove them entirely if desired. The **l2** regularization parameter will not reduce the coefficient of features to zero, so we will observe that they all have non-zero values.

The following code shows how to instantiate the models using these regularization techniques:

```
model_l1 = LogisticRegressionCV(Cs=Cs, penalty='l1', \
                                cv=10, solver='liblinear', \
                                random_state=42)
model_l2 = LogisticRegressionCV(Cs=Cs, penalty='l2', \
                                cv=10, random_state=42)
```

The following code shows how to fit the models:

```
model_l1.fit(X_train, y_train['Revenue'])
model_l2.fit(X_train, y_train['Revenue'])
```

The same concepts in lasso and ridge regularization can be applied to ANNs. However, penalization occurs on the weight matrices rather than the coefficients. Dropout is another form of regularization that's used to prevent overfitting in ANNs. Dropout randomly selects nodes at each iteration and removes them, along with their connections, as shown in the following figure:

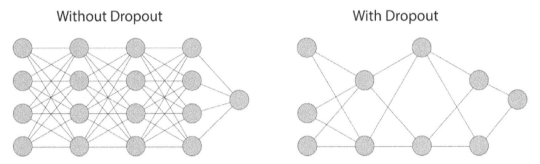

Figure 1.35: Dropout regularization in ANNs

CROSS-VALIDATION

Cross-validation is often used in conjunction with regularization to help tune hyperparameters. Take, for example, the **penalization** parameter in ridge and lasso regression, or the proportion of nodes to drop out at each iteration using the dropout technique with ANNs. How will you determine which parameter to use? One way is to run models for each value of the regularization parameter and evaluate them on the test set; however, using the test set often can introduce bias into the model.

One popular example of cross-validation is called k-fold cross-validation. This technique gives us the ability to test our model on unseen data while retaining a test set that we will use to test at the end. Using this method, the data is divided into **k** subsets. In each of the **k** iterations, **k-1** of the subsets are used as training data and the remaining subset is used as a validation set. This is repeated **k** times until all *k* subsets have been used as validation sets.

By using this technique, there is a significant reduction in bias, since most of the data is used for fitting. There is also a reduction in variation since most of the data is also used for validation. Typically, there are between **5** and **10** folds, and the technique can even be stratified, which is useful when there is a large imbalance of classes.

The following example shows **5-fold cross-validation** with **20%** of the data being held out as a test set. The remaining **80%** is separated into 5 folds. Four of those folds comprise the training data, and the remaining fold is the validation data. This is repeated a total of five times until every fold has been used once for validation:

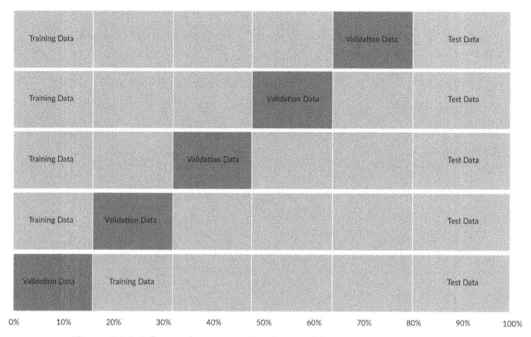

Figure 1.36: A figure demonstrating how 5-fold cross-validation works

ACTIVITY 1.01: ADDING REGULARIZATION TO THE MODEL

In this activity, we will utilize the same logistic regression model from the scikit-learn package. This time, however, we will add regularization to the model and search for the optimum regularization parameter—a process often called hyperparameter tuning. After training the models, we will test the predictions and compare the model evaluation metrics to those produced by the baseline model and the model without regularization.

The steps we will take are as follows:

1. Load in the feature and target datasets of the online shoppers purchasing intention dataset from `'../data/OSI_feats_e3.csv'` and `'../data/OSI_target_e2.csv'`.

2. Create **training** and **test** datasets for each of the **feature** and **target** datasets. The **training** datasets will be used to train on, and the models will be evaluated using the **test** datasets.

3. Instantiate a model instance of the **LogisticRegressionCV** class of scikit-learn's **linear_model** package.

4. Fit the model to the **training** data.

5. Make predictions on the **test** dataset using the trained model.

6. Evaluate the models by comparing how they scored against the **true** values using the evaluation metrics.

After implementing these steps, you should get the following expected output:

```
l1
Precision: 0.7300
Recall: 0.4078
fscore: 0.5233


l2
Precision: 0.7350
Recall: 0.4106
fscore: 0.5269
```

> **NOTE**
>
> The solution for this activity can be found on page 348.

This activity has taught us how to use **regularization** in **conjunction** with **cross-validation** to appropriately score a model. We have learned how to fit a model to data using regularization and cross-validation. Regularization is an important technique to use to ensure that models don't overfit the training data. Models that have been trained with regularization will perform better on new data, which is generally the goal of machine learning models—to predict a target when given new observations of the input data. Choosing the optimal regularization parameter may require iterating over a number of different choices.

Cross-validation is a technique that's used to determine which set of regularization parameters fit the data best. Cross-validation will train multiple models with different values for the regularization parameters on different cuts of the data. This technique ensures the best set of regularization parameters are chosen, without adding bias and minimizing variance.

SUMMARY

In this chapter, we covered how to prepare data and construct machine learning models. We achieved this by utilizing Python and libraries such as pandas and scikit-learn. We also used the algorithms in scikit-learn to build our machine learning models.

Then, we learned how to load data into Python, as well as how to manipulate data so that a machine learning model can be trained on the data. This involved converting all the columns into numerical data types. We also created a basic logistic regression classification model using scikit-learn algorithms. We divided the dataset into training and test datasets and fit the model to the training dataset. We evaluated the performance of the model on the test dataset using the model evaluation metrics, that is, accuracy, precision, recall, and fscore.

Finally, we iterated on this basic model by creating two models with different types of regularization for the model. We utilized cross-validation to determine the optimal parameter to use for the regularization parameter.

In the next chapter, we will use these same concepts to create the model using the Keras library. We will use the same dataset and attempt to predict the same target value for the same classification task. By doing so, we will learn how to use `regularization`, `cross-validation`, and `model evaluation metrics` when fitting our neural network to the data.

2

MACHINE LEARNING VERSUS DEEP LEARNING

OVERVIEW

In this chapter, we will begin creating Artificial Neural Networks (ANNs) using the Keras library. Before utilizing the library for modeling, we will get an introduction to the mathematics that comprise ANNs—understanding linear transformations and how they can be applied in Python. You'll build a firm grasp of the mathematics that make up ANNs. By the end of this chapter, we will have applied that knowledge by building a logistic regression model with Keras.

INTRODUCTION

In the previous chapter, we discussed some applications of machine learning and even built models with the scikit-learn Python package. The previous chapter covered how to preprocess real-world datasets so that they can be used for modeling. To do this, we converted all the variables into numerical data types and converted `categorical` variables into `dummy` variables. We used the `logistic regression` algorithm to classify users of a website by their purchase intention from the `online shoppers purchasing intention` dataset. We advanced our model-building skills by adding `regularization` to the dataset to improve the performance of our models.

In this chapter, we will continue learning how to build machine learning models and extend our knowledge so that we can build an **Artificial Neural Network** (**ANN**) with the Keras package. (Remember that **ANNs** represent a large class of machine learning algorithms that are so-called because their architecture resembles the neurons in the human brain.)

Keras is a machine learning library designed specifically for building neural networks. While scikit-learn's functionality spans a broader area of machine learning algorithms, the functionality of `scikit-learn` for neural networks is minimal.

ANNs can be used for the same machine learning tasks that other algorithms can perform, such as `logistic regression` for `classification` tasks, `linear regression` for `regression` problems, and `k-means` for `clustering`. Whenever we begin any machine learning problem, to determine what kind of task it is (`regression`, `classification`, or `clustering`), we need to ask the following questions:

- **What outcomes matter the most to me or my business?** For example, if you are predicting the value of stock market indices, you could predict whether the price is higher or lower than the previous time point (which would be a `classification` task) or you could predict the value itself (which would be a `regression` problem). Each may lead to a different subsequent action or trading strategy.

The following plot shows a **candlestick chart**. It describes the price movements in financial data and is depicting a stock price. The colors represent whether the stock price increased (green) or decreased (red) in value over each period, and each candlestick shows the open, close, high, and low values of the data—important pieces of information for stock prices.

> **NOTE**
>
> You can find the high-quality color images for this chapter at: https://packt.live/38nenXS.

One goal of modeling this data would be to predict what happens the following day. A **classification** task might predict a positive or negative change in the stock price and since there are only two possible values, this would be a binary classification task. Another option would be to predict the value of the stock the following day. Since the predicted value would be a **continuous** variable, this would be a **regression** task:

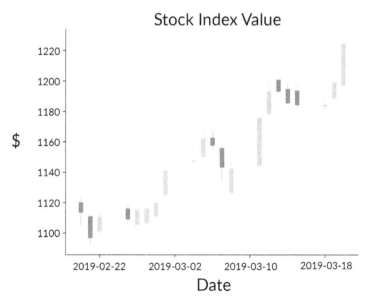

Figure 2.1: A candlestick chart indicating the movement of a stock index over the span of a month

- **Do we have the appropriately labeled data to train a model?** For a supervised learning task, we must have at least some labeled data in order to train a model. For example, if we want to build a model to classify images into dog images and cat images, we would need training data, the images themselves, and labels for the data indicating whether they are dog images or cat images. ANNs often need a lot of data. For image classification, this can be millions of images to develop accurate, robust models. This may be a determining factor when deciding which algorithm is appropriate for a given task.

ANNs are a type of machine learning algorithm that can be used to solve a task. They excel in certain aspects and have drawbacks in others, and these pros and cons should be considered before choosing this type of algorithm. Deep learning networks are distinguished from single-layer ANNs by their depth—the total number of hidden layers within the network.

So, deep learning is really just a specific subgroup of machine learning that relies on ANNs with multiple layers. We encounter the results of deep learning on a regular basis, whether it's in image classification models such as the friend recognition models that help tag friends in your Facebook photos, or the recommendation algorithms that help suggest your next favorite songs on Spotify. Deep learning models are becoming more prevalent over traditional machine learning models for a variety of reasons, including the growing sizes of unstructured data that deep learning models excel at and lower computational costs.

Choosing whether to use ANNs or traditional machine learning algorithms such as linear regression and decision trees for a particular task is a matter of experience and an understanding of the inner workings of the algorithm itself. As such, the benefits of using traditional machine learning algorithms or ANNs will be mentioned in the next section.

ADVANTAGES OF ANNS OVER TRADITIONAL MACHINE LEARNING ALGORITHMS

- **The best performance**: For any supervised learning task, the best models have been ANNs that are trained on a lot of data. For example, in classification tasks such as classifying images from the `ImageNet challenge` (a large-scale visual recognition challenge for classifying images into `1000 classes`), ANNs can attain greater accuracy than humans.

- **Scale effectively with data**: Traditional machine learning algorithms, such as `logistic regression` and `decision trees`, plateau in performance, whereas the ANN architecture is able to learn higher-level features—nonlinear combinations of the input features that may be important for classification or regression tasks. This allows ANNs to perform better when provided with large amounts of data - especially those ANNs with a deep architecture. For example, ANNs that perform well in the ImageNet challenge are provided with `14 million images` for training. The following figure shows the performance scaling with the amount of data for both deep learning algorithms and traditional machine learning algorithms:

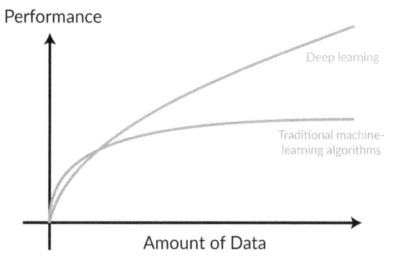

Figure 2.2: Performance scaling with the amount of data for both deep learning algorithms and traditional machine learning algorithms

- **No need for feature engineering**: ANNs are able to identify which features are important in modeling so that they are able to model directly from raw data. For example, in the binary classification of dog and cat images into their respective classes, there is no need to define features such as the color size or weight of the animal. The images themselves are sufficient for the ANN to successfully determine classification. In traditional machine learning algorithms, these features must be engineered in an iterative process that is manual and can be time-consuming.

- **Adaptable and transferable**: Weights and features that are learned from ANNs can be applied to similar tasks. In computer vision tasks, pre-trained classification models can be used as the starting points for building models for other classification tasks. For example, VGG-16 is a `16-layer deep learning model` that's used by **ImageNet** to classify `1000 random objects`. The weights that are learned in the model can be transferred to classify other objects in significantly less time.

However, there are some advantages of using traditional machine learning algorithms over ANNs, as explained in the following section.

ADVANTAGES OF TRADITIONAL MACHINE LEARNING ALGORITHMS OVER ANNS

- **Relatively good performance when the available training data is small**: In order to attain high performance, ANNs require a lot of data, and the deeper the network, the more data is required. With the increase in layers, the number of parameters that need to be learned also increases. This results in more time to train on the training data to reach the optimal parameter values. For example, `VGG-16` has over `138 million parameters` and required `14 million hand-labeled images` to train and learn all the parameters.

- **Cost-effective**: Both financially and computationally, deep networks can take a lot of computing power and time to train. This demands a lot of resources that may not be available to all. Moreover, these models are time-consuming to tune effectively and require a domain expert who's familiar with the inner workings of the model to achieve optimal performance.

- **Easy to interpret**: Many traditional machine learning models are easy to interpret. So, identifying which feature had the most predictive power in the model is straightforward. This can be incredibly useful when working with non-technical team members who wish to understand and interpret the results of the model. ANNs are considered more of a `black box`, in that while they are successful in classifying images and other tasks, the understanding behind how the predictions are made is unintuitive and buried in layers of computations. As such, interpreting the results requires more effort.

HIERARCHICAL DATA REPRESENTATION

One reason that ANNs are able to perform so well is that a large number of layers allows the network to learn representations of the data at many different levels. This is illustrated in the following diagram, in which the representation of an ANN being used to identify faces is shown. At lower levels of the model, simple features are learned, such as edges and gradients, as can be seen by looking at the features that were learned in the initial layers. As the model progresses, combinations of lower-level features activate to form face parts, and at later layers of the model, generic faces are learned. This is known as feature hierarchy and illustrates the power that this layered representation has for model building and interpretation.

Many examples of input for real-world applications of deep neural networks involve images, video, and natural language text. The feature hierarchy that is learned by deep neural networks allows them to discover latent structures within unlabeled, unstructured data, such as images, video, and natural language text, which makes them useful for processing real-world data—most often raw and unprocessed.

The following diagram shows an example of the learned representation of a deep learning model—lower features such as the **edges** and **gradients** activate together to form generic face shapes, which can be seen in the deeper layers:

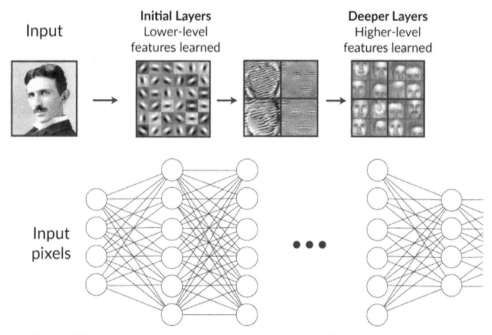

Figure 2.3: Learned representation at various parts of a deep learning model

Since deep neural networks have become more accessible, various companies have started exploiting their applications. The following are some examples of some companies that use ANNs:

- **Yelp**: Yelp uses deep neural networks to process, classify, and label their images more efficiently. Since photos are one important aspect of Yelp reviews, the company has placed an emphasis on classifying and categorizing them. This is achieved more efficiently with deep neural networks.

- **Clarifai**: This cloud-based company is able to classify images and videos using deep neural network-based models.

- **Enlitic**: This company uses deep neural networks to analyze medical image data such as X-rays or MRIs. The use of such networks in this application increases diagnostic accuracy and decreases diagnostic time and cost.

Now that we understand the potential applications of using ANNs, we can understand the mathematics behind how they work. While they may seem intimidating and complex, they can be broken down into a series of linear and nonlinear transformations, which themselves are simple to understand. An ANN is created by sequentially combining a series of linear and nonlinear transformations. The next section discusses the basic components and operations involved in linear transformations that comprise the mathematics of ANNs.

LINEAR TRANSFORMATIONS

In this section, we will introduce linear transformations. Linear transformations are the backbone of modeling with ANNs. In fact, all the processes of ANN modeling can be thought of as a series of linear transformations. The working components of linear transformations are scalars, vectors, matrices, and tensors. Operations such as **addition**, **transposition**, and **multiplication** are performed on these components.

SCALARS, VECTORS, MATRICES, AND TENSORS

Scalars, **vectors**, **matrices**, and **tensors** are the actual components of any deep learning model. Having a fundamental understanding of how to utilize these components, as well as the operations that can be performed on them, is key to understanding how ANNs operate. **Scalars**, **vectors**, and **matrices** are examples of the general entity known as a **tensor**, so the term **tensors** may be used throughout this chapter but may refer to any component. **Scalars**, **vectors**, and **matrices** refer to **tensors** with a specific number of dimensions.

The rank of a **tensor** is an attribute that determines the number of dimensions the **tensor** spans. The definitions of each are listed here:

- **Scalar**: They are single numbers and are an example of 0-order **tensors**. For instance, the temperature at any given point is a **scalar** field.

- **Vector**: Vectors are one-dimensional arrays of single numbers and are an example of first-order **tensors**. The **velocity** of a given object is an example of a **vector** field since it will have a speed in the **two (x,y)** or **three (x,y,z)** dimensions.

- **Matrix**: **Matrices** are rectangular arrays that span over two dimensions that consist of single numbers. They are an example of second-order **tensors**. An example of where **matrices** might be used is to store the **velocity** of a given object over time. One dimension of the **matrix** comprises the speed in the given directions, while the other **matrix** dimension is comprised of each given time point.

- **Tensor**: **Tensors** are the general entities that encapsulate **scalars**, **vectors**, and **matrices**. In general, the name is reserved for **tensors** of order **3** or more. An example of where **tensors** might be used is to store the **velocity** of many objects over time. One dimension of the **matrix** comprises the speed in the given directions, another **matrix** dimension is given for each given time point, and a third dimension describes the various objects.

The following diagram shows some examples of a **scalar**, a **vector**, a **matrix**, and a **three-dimensional tensor**:

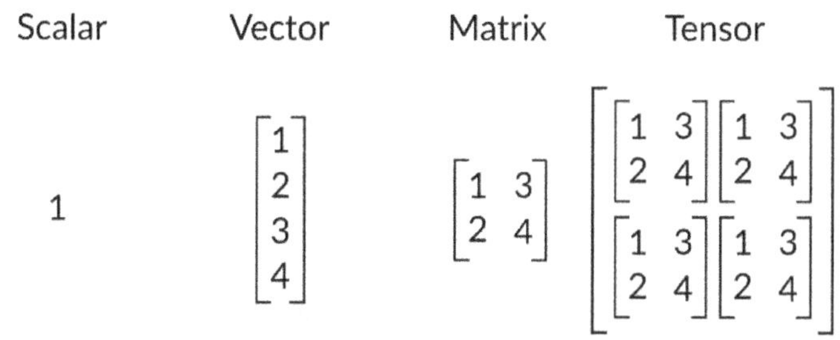

Figure 2.4: A visual representation of scalars, vectors, matrices, and tensors

TENSOR ADDITION

Tensors can be added together to create new **tensors**. We will use the example of matrices in this chapter, but this concept can be extended to **tensors** with any rank. **Matrices** may be added to **scalars**, **vectors**, and other **matrices** under certain conditions.

Two matrices may be added (or subtracted) together if they have the same shape. For such matrix-matrix addition, the resultant matrix is determined by the element-wise addition of the input matrices. The resultant matrix will, therefore, have the same shape as the two input matrices. We can define the matrix C = [c_{ij}] as the matrix sum **C = A + B**, where c_{ij} = a_{ij} + b_{ij} and each element in **C** is the sum of the same element in **A** and **B**. Matrix addition is commutative, which means that the order of **A** and **B** does not matter – **A + B = B + A**. Matrix addition is also associative, which means that the same result is achieved, even when the order of additions is different or even if the operation is applied more than once: **A + (B + C) = (A + B) + C**.

The same matrix addition principles apply for **scalars**, **vectors**, and **tensors**. An example of this is as follows:

$$A + B = \begin{bmatrix} 1 & 4 & 1 \\ 9 & 2 & 5 \\ 7 & 3 & 1 \end{bmatrix} + \begin{bmatrix} 5 & 2 & 4 \\ 7 & 4 & 2 \\ 2 & 3 & 8 \end{bmatrix} = \begin{bmatrix} 6 & 6 & 5 \\ 16 & 6 & 7 \\ 9 & 6 & 9 \end{bmatrix} = C$$

Figure 2.5: An example of matrix-matrix addition

Scalars can also be added to **matrices**. Here, each element of the **matrix** is added to the **scalar** individually, as is shown in the below figure:

$$A + 4 = \begin{bmatrix} 1 & 4 & 1 \\ 9 & 2 & 5 \\ 7 & 3 & 1 \end{bmatrix} + 4 = \begin{bmatrix} 5 & 8 & 5 \\ 13 & 6 & 9 \\ 11 & 7 & 5 \end{bmatrix} = B$$

Figure 2.6: An example of matrix-scalar addition

It is possible to add vectors to matrices if the number of columns between the two matches each other. This is known as broadcasting.

EXERCISE 2.01: PERFORMING VARIOUS OPERATIONS WITH VECTORS, MATRICES, AND TENSORS

> **NOTE**
>
> For the exercises and activities within this chapter, you will need to have Python 3.7, Jupyter, and NumPy installed on your system. All the exercises and activities will be primarily developed in Jupyter notebooks. It is recommended to keep a separate notebook for different assignments unless advised not to. Use the following link to download them from this book's GitHub repository: https://packt.live/2vpc9rO.

In this exercise, we are going to demonstrate how to create and work with **vectors**, **matrices**, and **tensors** within Python. We will assume that you have some familiarity with scalars. This can all be achieved with the NumPy library using the **array** and **matrix** functions. Tensors of any rank can be created with the NumPy **array** function.

Before you begin, you should set up the files and folders for this chapter in your working directory using a similar structure and naming convention as you did in the previous chapter. You can verify your folder structure by comparing it to the GitHub repository, linked above.

Follow these steps to perform this exercise:

1. Open Jupyter Notebook to implement this exercise. Import the necessary dependency. Create a **one-dimensional array**, or a **vector**, as follows:

```
import numpy as np
vec1 = np.array([1, 2, 3, 4, 5, 6, 7, 8, 9, 10])
vec1
```

The preceding code produces the following output:

```
array([ 1, 2, 3, 4, 5, 6, 7, 8, 9, 10])
```

2. Create a **two-dimensional array**, or **matrix**, with the **array** function:

    ```
    mat1 = np.array([[1, 2, 3], [4, 5, 6], [7, 8, 9], [10, 11, 12]])
    mat1
    ```

    The preceding code produces the following output:

    ```
    array([[ 1,  2,  3],
           [ 4,  5,  6],
           [ 7,  8,  9],
           [10, 11, 12]])
    ```

3. Use the **matrix** function to create matrices, which will show a similar output:

    ```
    mat2 = np.matrix([[1, 2, 3], [4, 5, 6], \
                      [7, 8, 9], [10, 11, 12]])
    ```

4. Create a **three-dimensional array**, or **tensor**, using the **array** function:

    ```
    ten1 = np.array([[[1, 2, 3], [4, 5, 6]], \
                     [[7, 8, 9], [10, 11, 12]]])
    ten1
    ```

    The preceding code produces the following output:

    ```
    array([[[ 1,  2,  3],
            [ 4,  5,  6],
           [[ 7,  8,  9],
            [10, 11, 12]]])
    ```

5. Determining the **shape** of a given **vector**, **matrix**, or **tensor** is important since certain operations, such as **addition** and **multiplication**, can only be applied to components of certain shapes. The shape of an n-dimensional array can be determined using the **shape** method. Write the following code to determine the **shape** of **vec1**:

    ```
    vec1.shape
    ```

    The preceding code produces the following output:

    ```
    (10, )
    ```

6. Write the following code to determine the **shape** of **mat1**:

```
mat1.shape
```

The preceding code produces the following output:

```
(4, 3)
```

7. Write the following code to determine the **shape** of **ten1**:

```
ten1.shape
```

The preceding code produces the following output:

```
(2, 2, 3)
```

8. Create a **matrix** with **four rows** and **three columns** with whichever numbers you like. Print the resulting matrix to verify its **shape**:

```
mat1 = np.matrix([[1, 2, 3], [4, 5, 6], [7, 8, 9], [10, 11, 12]])
mat1
```

The preceding code produces the following output:

```
matrix([[ 1,  2,  3],
        [ 4,  5,  6],
        [ 7,  8,  9],
        [10, 11, 12]])
```

9. Create another matrix with **four rows** and **three columns** with whichever numbers you like. Print the resulting matrix to verify its **shape**:

```
mat2 = np.matrix([[2, 1, 4], [4, 1, 7], [4, 2, 9], [5, 21, 1]])
mat2
```

The preceding code produces the following output:

```
matrix([[ 2,  1,  4],
        [ 4,  1,  7],
        [ 4,  2,  9],
        [ 5, 21,  1]])
```

10. Add **matrix 1** and **matrix 2**:

```
mat3 = mat1 + mat2
mat3
```

The preceding code produces the following output:

```
matrix([[  3,  3,  7],
        [  8,  6, 13],
        [ 11, 10, 18],
        [ 15, 32, 13]])
```

11. Add **scalars** to the **arrays** with the following code:

```
mat1 + 4
```

The preceding code produces the following output:

```
matrix([[  5,  6,  7],
        [  8,  9, 10],
        [ 11, 12, 13],
        [ 14, 15, 16]])
```

In this exercise, we learned how to perform various operations with **vectors**, **matrices**, and **tensors**. We also learned how to determine the **shape** of the **matrix**.

> **NOTE**
>
> To access the source code for this specific section, please refer to https://packt.live/2NNQ7VA.
>
> You can also run this example online at https://packt.live/3eUDtQA.

RESHAPING

A **tensor** of any size can be reshaped as long as the number of total elements remains the same. For example, a **(4x3) matrix** can be reshaped into a **(6x2) matrix** since they both have a total of **12** elements. The **rank**, or **number of dimensions**, can also be changed in the **reshaping** process. For example, a **(4x3) matrix** can be reshaped into a **(3x2x2) tensor**. Here, the **rank** has changed from **2** to **3**. The **(4x3) matrix** can also be reshaped into a **(12x1) vector**, in which the **rank** has changed from **2** to **1**.

The following diagram illustrates tensor reshaping—on the left is a tensor with **shape (4x1x3)**, which can be reshaped to a tensor of **shape (4x3)**. Here, the number of elements **(12)** has remained constant, though the **shape** and **rank** of the tensor have changed:

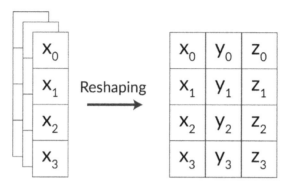

Figure 2.7: Visual representation of reshaping a (4x1x3) tensor into a (4x3) tensor

MATRIX TRANSPOSITION

The **transpose** of a matrix is an operator that flips the matrix over its diagonal. When this occurs, the rows become the columns and vice versa. The transpose operation is usually denoted as a **T** superscript upon the matrix. Tensors of any rank can also be transposed:

$$\begin{bmatrix} & m & \\ & & \\ n & & \\ & & \end{bmatrix}^T = \begin{bmatrix} & n & \\ m & & \\ & & \end{bmatrix}$$

Figure 2.8: A visual representation of matrix transposition

The following figure shows the matrix transposition properties of matrices **A** and **B**:

$$\left(A^T\right)^T = A$$
$$(A + B)^T = A^T + B^T$$
$$(AB)^T = B^T A^T$$
$$(A_1 A_2 \ldots A_k)^T = A_k^T \ldots A_2^T A_1^T$$
$$\left(A^{-1}\right)^T = \left(A^T\right)^{-1}$$

Figure 2.9: Matrix transposition properties where A and B are matrices

A square matrix (that is, a matrix with an equivalent number of rows and columns) is said to be symmetrical if the transpose of a matrix is equivalent to the original matrix.

EXERCISE 2.02: MATRIX RESHAPING AND TRANSPOSITION

In this exercise, we are going to demonstrate how to reshape and transpose matrices. This will become important since some operations can only be applied to components if certain tensor dimensions match. For example, tensor multiplication can only be applied if the inner dimensions of the two tensors match. Reshaping or transposing tensors is one way to modify the dimensions of the tensor to ensure that certain operations can be applied. Follow these steps to complete this exercise:

1. Open a Jupyter notebook from the start menu to implement this exercise. Create a **two-dimensional array** with **four rows** and **three columns**, as follows:

```
import numpy as np
mat1 = np.array([[1, 2, 3], [4, 5, 6], [7, 8, 9], [10, 11, 12]])
mat1
```

This gives the following output:

```
array([[ 1,  2,  3],
       [ 4,  5,  6],
       [ 7,  8,  9],
       [10, 11, 12]])
```

We can confirm its shape by looking at the shape of the matrix:

```
mat1.shape
```

The output is as follows:

```
(4, 3)
```

2. Reshape the array so that it has **three rows** and **four columns** instead, as follows:

```
mat2 = np.reshape(mat1, [3,4])
mat2
```

The preceding code produces the following output:

```
array([[ 1, 2, 3, 4],
       [ 5, 6, 7, 8],
       [ 9, 10, 11, 12]])
```

3. Confirm this by printing the **shape** of the array:

```
mat2.shape
```

The preceding code produces the following output:

```
(3, 4)
```

4. Reshape the matrix into a **three-dimensional array**, as follows:

```
mat3 = np.reshape(mat1, [3,2,2])
mat3
```

The preceding code produces the following output:

```
array([[[ 1,  2],
        [ 3,  4]],

       [[ 5,  6],
        [ 7,  8]],

       [[ 9, 10],
        [ 11, 12]]])
```

5. Print the **shape** of the array to confirm its dimensions:

```
mat3.shape
```

The preceding code produces the following output:

```
(3, 2, 2)
```

6. Reshape the matrix into a **one-dimensional array**, as follows:

```
mat4 = np.reshape(mat1, [12])
mat4
```

The preceding code produces the following output:

```
array([ 1,  2,  3,  4,  5,  6,  7,  8,  9, 10, 11, 12])
```

7. Confirm this by printing the **shape** of the array:

```
mat4.shape
```

The preceding code produces the following output:

```
(12, )
```

8. Taking the transpose of an array will flip it across its diagonal. For a one-dimensional array, a row-vector will be converted into a column vector and vice versa. For a two-dimensional array or matrix, each row becomes a column and vice versa. Call the transpose of an array using the **T** method:

```
mat = np.matrix([[1, 2, 3], [4, 5, 6], [7, 8, 9], [10, 11, 12]])
mat.T
```

The following figure shows the output of the preceding code:

$$mat= \quad \begin{matrix} \texttt{matrix([[1,} & \texttt{2,} & \texttt{3],} \\ \texttt{[4,} & \texttt{5,} & \texttt{6],} \\ \texttt{[7,} & \texttt{8,} & \texttt{9],} \\ \texttt{[10,} & \texttt{11,} & \texttt{12]])} \end{matrix} \qquad mat^T= \quad \begin{matrix} \texttt{matrix([[1,} & \texttt{4,} & \texttt{7,} & \texttt{10],} \\ \texttt{[2,} & \texttt{5,} & \texttt{8,} & \texttt{11],} \\ \texttt{[3,} & \texttt{6,} & \texttt{9,} & \texttt{12]])} \end{matrix}$$

Figure 2.10: Visual demonstration of the transpose function

9. Check the **shape** of the matrix and its transpose to verify that the dimensions have changed:

```
mat.shape
```

The preceding code produces the following output:

```
(4, 3)
```

10. Check the **shape** of the transposed matrix:

```
mat.T.shape
```

The preceding code produces the following output:

```
(3, 4)
```

11. Verify the matrix elements do not match when a matrix is reshaped, and a matrix is transposed:

```
np.reshape(mat1, [3,4]) == mat1.T
```

The preceding code produces the following output:

```
array([[ True, False, False, False],
       [False, False, False, False],
       [False, False, False, True]], dtype = bool)
```

Here, we can see that only the first and last elements match.

In this section, we introduced some of the basic components of linear algebra, including scalars, vectors, matrices, and tensors. We also covered some basic manipulation of linear algebra components, such as addition, transposition, and reshaping. By doing so, we learned how to put these concepts into action by using functions in the **NumPy** library to perform these operations.

> **NOTE**
>
> To access the source code for this specific section, please refer to https://packt.live/3gqBlR0.
>
> You can also run this example online at https://packt.live/3eYCChD.

In the next section, we will extend our understanding of linear transformations by covering one of the most important transformations related to ANNs— matrix multiplication.

MATRIX MULTIPLICATION

Matrix multiplication is fundamental to neural network operations. While the rules for addition are simple and intuitive, the rules for multiplication for matrices and tensors are more complex. Matrix multiplication involves more than simple element-wise multiplication of the elements. Instead, a more complicated procedure is implemented that involves the entire row of one matrix and an entire column of the other. In this section, we will explain how multiplication works for two-dimensional tensors or matrices; however, tensors of higher orders can also be multiplied.

Given a matrix, $A = [a_{ij}]_{m \times n}$, and another matrix, $B = [b_{ij}]_{n \times p}$, the product of the two matrices is $C = AB = [C_{ij}]_{m \times p}$, and each element, c_{ij}, is defined element-wise as $c_{ij} = \sum_{k=1}^{n} a_{ik} b_{kj}$. Note that the shape of the resultant matrix is the same as the outer dimensions of the matrix product or the number of rows of the first matrix and the number of columns of the second matrix. For the multiplication to work, the inner dimensions of the matrix product must match, or the number of columns of the first matrix and the number of columns of the second matrix.

The concept of inner and outer dimensions of matrix multiplication can be seen in the following figure:

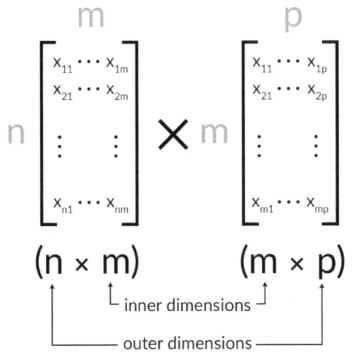

Figure 2.11: A visual representation of the inner and outer dimensions in matrix multiplication

Unlike matrix addition, matrix multiplication is not commutative, which means that the order of the matrices in the product matters:

$$AB \neq BA$$

Figure 2.12: Matrix multiplication is non-commutative

For example, let's say we have the following two matrices:

$$A = \begin{bmatrix} 2 & 5 & 1 \\ 7 & 3 & 6 \end{bmatrix}, B = \begin{bmatrix} 1 & 8 \\ 9 & 4 \\ 3 & 5 \end{bmatrix}$$

Figure 2.13: Two matrices, A and B

One way to construct the product is to have matrix **A** first, multiplied by **B**:

$$AB = \begin{bmatrix} 2 \times 1 + 5 \times 9 + 1 \times 3 & 2 \times 8 + 5 \times 4 + 1 \times 5 \\ 7 \times 1 + 3 \times 9 + 6 \times 3 & 7 \times 8 + 3 \times 4 + 6 \times 5 \end{bmatrix} = \begin{bmatrix} 50 & 41 \\ 52 & 98 \end{bmatrix}$$

Figure 2.14: Visual representation of matrix A multiplied by B

This results in a **2x2** matrix. Another way to construct the product is to have **B** first, multiplied by **A**:

$$BA = \begin{bmatrix} 1 \times 2 + 8 \times 7 & 1 \times 5 + 8 \times 3 & 1 \times 1 + 8 \times 6 \\ 9 \times 2 + 4 \times 7 & 9 \times 5 + 4 \times 3 & 9 \times 1 + 4 \times 6 \\ 3 \times 2 + 5 \times 7 & 3 \times 5 + 5 \times 3 & 3 \times 1 + 5 \times 6 \end{bmatrix} = \begin{bmatrix} 58 & 29 & 49 \\ 46 & 57 & 33 \\ 41 & 30 & 33 \end{bmatrix}$$

Figure 2.15: Visual representation of matrix B multiplied by A

Here, we can see that the matrix that was formed from the product **BA** is a **3x3** matrix and is very different from the matrix that was formed from the product **AB**.

Scalar-matrix multiplication is much more straightforward and is simply the product of every element in the matrix multiplied by the scalar so that $\lambda A = [\lambda a_{ij}]_{m \times n}$, where λ is a scalar and **A** is a matrix.

In the following exercise, we will put our understanding into practice by performing matrix multiplication in Python utilizing the **NumPy** library.

EXERCISE 2.03: MATRIX MULTIPLICATION

In this exercise, we are going to demonstrate how to multiply matrices together. Follow these steps to complete this exercise:

1. Open a Jupyter notebook from the start menu to implement this exercise.

    To demonstrate the fundamentals of matrix multiplication, begin with two matrices of the same shape:

```
import numpy as np
mat1 = np.array([[1, 2, 3], [4, 5, 6], \
                [7, 8, 9], [10, 11, 12]])
mat2 = np.array([[2, 1, 4], [4, 1, 7], \
                [4, 2, 9], [5, 21, 1]])
```

2. Since both matrices have the same shape and they are not square, they cannot be multiplied as is, otherwise, the inner dimensions of the product won't match. One way we could resolve this is to take the transpose of one of the matrices; then, we would be able to perform the multiplication. Take the transpose of the second matrix, which would mean that a **(4x3) matrix** is multiplied by a **(3x4) matrix**. The result would be a **(4x4) matrix**. Perform the multiplication using the **dot** method:

```
mat1.dot(mat2.T)
```

The preceding code produces the following output:

```
array([[ 16, 27, 35, 50],
       [ 37, 63, 80, 131],
       [ 58, 99, 125, 212],
       [ 79, 135, 170, 293]])
```

3. Take the transpose of the first matrix, which would mean that a **(3x4) matrix** is multiplied by a **(4x3) matrix**. The result would be a **(3x3) matrix**:

```
mat1.T.dot(mat2)
```

The preceding code produces the following output:

```
array([[ 96, 229, 105],
       [ 111, 254, 126],
       [ 126, 279, 147]])
```

4. Reshape one of the arrays to make sure the inner dimension of the matrix multiplication matches. For example, we can reshape the first array to make it a **(3x4) matrix** instead of transposing. Note that the result is not the same as it is when transposing:

```
np.reshape(mat1, [3,4]).dot(mat2)
```

The preceding code produces the following output:

```
array([[ 42, 93, 49],
       [ 102, 193, 133],
       [ 162, 293, 217]])
```

In this exercise, we learned how to multiply two matrices together. The same concept can be applied to tensors of all ranks, not just second-order tensors. Tensors of different ranks can even be multiplied together if their inner dimensions match.

> **NOTE**
>
> To access the source code for this specific section, please refer to https://packt.live/38p0RD7.
>
> You can also run this example online at https://packt.live/2VYl1xZ.

The next exercise demonstrates how to multiply three-dimensional tensors together.

EXERCISE 2.04: TENSOR MULTIPLICATION

In this exercise, we are going to apply our knowledge of matrix multiplication to higher-order tensors. Follow these steps to complete this exercise:

1. Open a Jupyter notebook from the start menu to implement this exercise. Begin by creating a three-dimensional tensor using the NumPy library and the **array** function. Import all the necessary dependencies:

```
import numpy as np
mat1 = np.array([[[1, 2, 3], [4, 5, 6]], [[1, 2, 3], [4, 5, 6]]])
mat1
```

The preceding code produces the following output:

```
array([[[ 1,  2,  3],
        [ 4,  5,  6],

       [[ 1,  2,  3],
        [ 4,  5,  6]]])
```

2. Confirm the shape using the **shape** method:

```
mat1.shape
```

This tensor has the shape (2x2x3).

3. Create a new **three-dimensional tensor** that we will be able to multiply the tensor by. Take the transpose of the original matrix:

```
mat2 = mat1.T
mat2
```

The preceding code produces the following output:

```
array([[[ 1,  1],
        [ 4,  4]],

       [[ 2,  2],
        [ 5,  5]],

       [[ 3,  3],
        [ 6,  6]]])
```

4. Confirm the shape using the **shape** method:

```
mat2.shape
```

This tensor has the shape (3x2x2).

5. Take the **dot** product of the **two matrices**, as follows:

```
mat3 = mat2.dot(mat1)
mat3
```

The preceding code produces the following output:

```
array([[[[ 5,  7,  9],
         [ 5,  7,  9]],

        [[ 20, 28, 36],
         [ 20, 28, 36]]],


       [[[ 10, 14, 18],
         [ 10, 14, 18]],

        [[ 25, 35, 45],
         [ 25, 35, 45]]],


       [[[ 15, 21, 27],
         [ 15, 21, 27]],

        [[ 30, 42, 54],
         [ 30, 42, 54]]]])
```

6. Look at the **shape** of this resultant tensor:

```
mat3.shape
```

The preceding code produces the following output:

```
(3, 2, 2, 3)
```

Now, we have a four-dimensional tensor.

In this exercise, we learned how to perform matrix multiplication using the NumPy library in Python. While we do not have to perform matrix multiplication directly when we create ANNs with Keras, it is still useful to understand the underlying mathematics.

> **NOTE**
>
> To access the source code for this specific section, please refer to https://packt.live/31G1rLn.
>
> You can also run this example online at https://packt.live/2AriZjn.

INTRODUCTION TO KERAS

Building ANNs involves creating layers of nodes. Each node can be thought of as a tensor of weights that are learned in the training process. Once the ANN has been fitted to the data, a prediction is made by multiplying the input data by the weight matrices layer by layer, applying any other linear transformation when needed, such as activation functions, until the final output layer is reached. The size of each weight tensor is determined by the size of the shape of the input nodes and the shape of the output nodes. For example, in a single-layer ANN, the size of our single hidden layer can be thought of as follows:

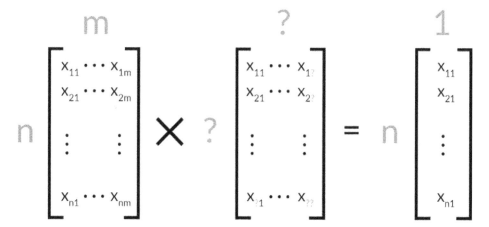

Figure 2.16: Solving the dimensions of the hidden layer of a single-layer ANN

If the input matrix of features has **n** rows, or observations, and **m** columns, or features, and we want our predicted target to have **n** rows (one for each observation) and one column (the predicted value), we can determine the size of our hidden layer by what is needed to make the matrix multiplication valid. Here is the representation of a single-layer ANN:

$$A_{n \times m} B_{? \times ?} = C_{n \times 1}$$

Figure 2.17: Representation of a single-layer ANN

Here, we can determine that the weight matrix will be of size (**mx1**) to ensure the matrix multiplication is valid.

If we have more than one hidden layer in an ANN, then we have much more freedom with the size of these weight matrices. In fact, the possibilities are endless, depending on how many layers there are and how many nodes we want in each layer. In practice, however, certain architecture designs work better than others, as we will be learning throughout this book.

In general, Keras abstracts much of the linear algebra out of building neural networks so that users can focus on designing the architecture. For most networks, only the input size, output size, and the number of nodes in each hidden layer are needed to create networks in Keras.

The simplest model structure in Keras is the **Sequential** model, which can be imported from **keras.models**. The model of the **Sequential** class describes an ANN that consists of a linear stack of layers. A **Sequential** model can be instantiated as follows:

```
from keras.models import Sequential
model = Sequential()
```

Layers can be added to this model instance to create the structure of the model.

> **NOTE**
>
> Before initializing your model, it is helpful to set a seed using the **seed** function in NumPy's random library and the **set_seed** function from TensorFlow's random library.

LAYER TYPES

The notion of layers is part of the Keras core API. A layer can be thought of as a composition of nodes, and at each node, a set of computations happen. In Keras, all the nodes of a layer can be initialized by simply initializing the layer itself. The individual operation of a generalized layer node can be seen in the following diagram. At each node, the input data is multiplied by a set of weights using matrix multiplication, as we learned earlier in this chapter. The sum of the product between the weights and the input is applied, which may or may not include a bias, as shown by the input node equal to **1** in the following diagram. Further functions may be applied to the output of this matrix multiplication, such as activation functions:

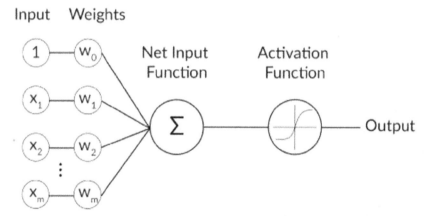

Figure 2.18: A depiction of a layer node

Some common layer types in Keras are as follows:

- **Dense**: This is a fully connected layer in which all the nodes of the layer are directly connected to all the inputs and all the outputs. ANNs for classification or regression tasks on tabular data usually have a large percentage of their layers with this type in the architecture.

- **Convolutional**: This layer type creates a convolutional kernel that is convolved with the input layer to produce a tensor of outputs. This convolution can occur in one or multiple dimensions. ANNs for the classification of images usually feature one or more convolutional layers in their architecture.

- **Pooling**: This type of layer is used to reduce the dimensionality of an input layer. Common types of pooling include max pooling, in which the maximum value of a given window is passed through to the output, or average pooling, in which the average value of a window is passed through. These layers are often used in conjunction with a convolutional layer, and their purpose is to reduce the dimensions of the subsequent layers, allowing for fewer training parameters to be learned with little information loss.

- **Recurrent**: Recurrent layers learn patterns from sequences, so each output is dependent on the results from the previous step. ANNs that model sequential data such as natural language or time-series data often feature one or more recurrent layer types.

There are other layer types in Keras; however, these are the most common types when it comes to building models using Keras.

Let's demonstrate how to add layers to a model by instantiating a model of the **Sequential** class and adding a **Dense** layer to the model. Successive layers can be added to the model in the order in which we wish the computation to be performed and can be imported from **keras.layers**. The number of units, or nodes, needs to be specified. This value will also determine the shape of the result from the layer. A **Dense** layer can be added to a **Sequential** model in the following way:

```
from keras.layers import Dense
from keras.models import Sequential
input_shape = 20
units = 1
model.add(Dense(units, input_dim=input_shape))
```

> **NOTE**
>
> After the first layer, the input dimension does not need to be specified since it is determined from the previous layer.

ACTIVATION FUNCTIONS

An activation function is generally applied to the output of a node to limit or bound its value. The value from each node is unbounded and may have any value, from negative to positive infinity. These can be troublesome within neural networks where the values of the weights and losses have been calculated and can head toward infinity and produce unusable results. Activation functions can help in this regard by bounding the value. Often, these activation functions push the value to two limits. Activation functions are also useful for deciding whether the node should be "fired" or not. Common activation functions are as follows:

- The **Step** function: The value is nonzero if it is above a certain threshold; otherwise, it is zero.

- The **Linear** function: $A(x) = cx$, which is a scalar multiplication of the input value.

- The **Sigmoid** function: $A(x) = \dfrac{1}{1+e^{-x}}$, such as a smoothed-out step function with smooth gradients. This activation function is useful for classification since the values are bound from zero to one.

- The **Tanh** function: $A(x) = \tanh(x) = \dfrac{2}{1+e^{-2x}} - 1$, which is a scaled version of the sigmoid with steeper gradients around **x=0**.

- The **ReLU** function: $A(x) = x,\ x > 0$, otherwise 0.

Now that we have looked at some of the main components, we can begin to see how we might create useful neural networks out of these components. In fact, we can create a logistic regression model with all the concepts we have learned about in this chapter. A logistic regression model operates by taking the sum of the product of an input and a set of learned weights, followed by the output being passed through a logistic function. This can be achieved with a single-layer neural network with a sigmoid activation function.

Activation functions can be added to models in the same manner that layers are added to models. The activation function will be applied to the output of the previous step in the model. A **tanh** activation function can be added to a **Sequential** model as follows:

```
from keras.layers import Dense, Activation
from keras.models import Sequential
input_shape = 20
units = 1
model.add(Dense(units, input_dim=input_shape))
model.add(Activation('tanh'))
```

> **NOTE**
>
> Activation functions can also be added to a model by including them as an argument when defining the layers.

MODEL FITTING

Once a model's architecture has been created, the model must be compiled. The compilation process configures all the learning parameters, including which optimizer to use, the loss function to minimize, as well as optional metrics, such as accuracy, to calculate at various stages of the model training. Models are compiled using the **compile** method, as follows:

```
model.compile(optimizer='adam', loss='binary_crossentropy', \
              metrics=['accuracy'])
```

After the model has been compiled, it is ready to be fit to the training data. This is achieved with an instantiated model using the **fit** method. Useful arguments when using the **fit** method are as follows:

- **X**: The array of the training feature data to fit the data to.

- **y**: The array of the training target data.

- **epochs**: The number of epochs to run the model for. An epoch is an iteration over the entire training dataset.

- **batch_size**: The number of training data samples to use per gradient update.

- **validation_split**: The proportion of the training data to be used for validation that is evaluated after each epoch.

- **shuffle**: Indicates whether to shuffle the training data before each epoch.

The **fit** method can be used on a model in the following way:

```
history = model.fit(x=X_train, y=y_train['y'], \
                    epochs=10, batch_size=32, \
                    validation_split=0.2, shuffle=False)
```

It is beneficial to save the output of calling the **fit** method of the model since it contains information on the model's performance throughout training, including the loss, which is evaluated after each epoch. If a validation split is defined, the loss is evaluated after each epoch on the validation split. Likewise, if any metrics are defined in training, they are also calculated after each epoch. It is useful to plot such loss and evaluation metrics to determine model performance as a function of the epoch. The model's loss as a function of the epoch can be visualized as follows:

```
import matplotlib.pyplot as plt
%matplotlib inline

plt.plot(history.history['loss'])
plt.show()
```

Keras models can be evaluated by utilizing the **evaluate** method of the model instance. This method returns the loss and any metrics that were passed to the model for training. The method can be called as follows when evaluating an out-of-sample test dataset:

```
test_loss = model.evaluate(X_test, y_test['y'])
```

These model-fitting steps represent the basic steps that need to be followed to build, train, and evaluate models using the Keras package. From here, there are an infinite number of ways to build and evaluate a model, depending on the task you wish to accomplish. In the following activity, we will create an ANN to perform the same task that we completed in *Chapter 1, Introduction to Machine Learning with Keras.* In fact, we will recreate the logistic regression algorithm with ANNs. As such, we expect there to be similar performance between the two models.

ACTIVITY 2.01: CREATING A LOGISTIC REGRESSION MODEL USING KERAS

In this activity, we are going to create a basic model using the Keras library. We will perform the same classification task that we did in *Chapter 1, Introduction to Machine Learning with Keras.* We will use the same online shopping purchasing intention dataset and attempt to predict the same variable.

In the previous chapter, we used a logistic regression model to predict whether a user would purchase a product from a website when given various attributes about the online session's behavior and the attributes of the web page. In this activity, we will introduce the Keras library, though we'll continue to utilize the libraries we introduced previously, such as **pandas**, for easily loading in the data, and **sklearn**, for any data preprocessing and model evaluation metrics.

> **NOTE**
>
> Preprocessed datasets have been provided for you to use for this activity. You can download them from https://packt.live/2ApIBwT.

The steps to complete this activity are as follows:

1. Load in the processed feature and target datasets.

2. Split the training and target data into training and test datasets. The model will be fit to the training dataset and the test dataset will be used to evaluate the model.

3. Instantiate a model of the **Sequential** class from the **keras.models** library.

4. Add a single layer of the **Dense** class from the **keras.layers** package to the model instance. The number of nodes should be equal to the number of features in the feature dataset.

5. Add a sigmoid activation function to the model.

6. Compile the model instance by specifying the optimizer to use, the loss metric to evaluate, and any other metrics to evaluate after each epoch.

7. Fit the model to the training data, specifying the number of epochs to run for and the validation split to use.

8. Plot the loss and other evaluation metrics with respect to the epoch that will be evaluated on the training and validation datasets.

9. Evaluate the loss and other evaluation metrics on the test dataset.

After implementing these steps, you should get the following expected output:

```
2466/2466 [==============================] - 0s 15us/step
The loss on the test set is 0.3632 and the accuracy is 86.902%
```

> **NOTE**
>
> The solution for this activity can be found on page 356.

In this activity, we looked at some of the fundamental concepts of creating ANNs in Keras, including various layer types and activation functions. We used these components to create a simple logistic regression model using a package that gives us similar results to the logistic regression model we used in *Chapter 1, Introduction to Machine Learning with Keras*. We learned how to build the model with the Keras library, train the model with a real-world dataset, and evaluate the performance of the model on a test dataset to provide an unbiased evaluation of the performance of the model.

SUMMARY

In this chapter, we covered the various types of linear algebra components and operations that pertain to machine learning. These components include scalars, vectors, matrices, and tensors. The operations that were applied to these tensors included addition, transposition, and multiplication—all of which are fundamental for understanding the underlying mathematics of ANNs.

We also learned some of the basics of the Keras package, including the mathematics that occurs at each node. We replicated the model from the previous chapter, in which we built a logistic regression model to predict the same target from the online shopping purchasing intention dataset. However, in this chapter, we used the Keras library to create the model using an ANN instead of the scikit-learn logistic regression model. We achieved a similar level of accuracy using ANNs.

The upcoming chapters of this book will use the same concepts we learned about in this chapter; however, we will continue building ANNs with the Keras package. We will extend our ANNs to more than a single layer by creating models that have multiple hidden layers. By adding multiple hidden layers to our ANNs, we will put the "deep" into "deep learning". We will also tackle the issues of underfitting and overfitting since they are related to training models with ANNs.

3

DEEP LEARNING WITH KERAS

OVERVIEW

In this chapter, you will experiment with different neural network architectures. You will create Keras sequential models—building single-layer and multi-layer models—and evaluate the performance of trained models. Networks of different architectures will help you understand overfitting and underfitting. By the end of this chapter, you will have explored early stopping that can be used to combat overfitting to the training data.

INTRODUCTION

In the previous chapter, you learned about the mathematics of neural networks, including **linear transformations** with **scalars**, **vectors**, **matrices**, and **tensors**. Then, you implemented your first neural network using Keras by building a logistic regression model to classify users of a website into those who will purchase from the website and those who will not.

In this chapter, you will extend your knowledge of building neural networks using Keras. This chapter covers the basics of deep learning and will provide you with the necessary foundations so that you can build highly complex neural network architectures. We will start by extending the **logistic regression** model to a simple single-layer neural network and then proceed to more complicated neural networks with multiple hidden layers.

In this process, you will learn about the underlying basic concepts of neural networks, including forward propagation for making predictions, computing loss, backpropagation for computing derivatives of loss with respect to model parameters, and, finally, gradient descent for learning about optimal parameters for the model. You will also learn about the various choices that are available so that you can build and train a neural network in terms of **activation functions**, **loss functions**, and **optimizers**.

Furthermore, you will learn how to evaluate your model while understanding issues such as **overfitting** and **underfitting**, all while looking at how they can impact the performance of your model and how to detect them. You will learn about the drawbacks of evaluating a model on the same dataset that's used for training, as well as the alternative approach of holding back a part of the available dataset for evaluation purposes. Subsequently, you will learn how to compare the model error rate on each of these two subsets of the dataset that can be used to detect problems such as high bias and high variance in the model. Lastly, you will learn about a technique called **early stopping** to reduce overfitting, which is again based on comparing the model's error rate to the two subsets of the dataset.

BUILDING YOUR FIRST NEURAL NETWORK

In this section, you will learn about the representations and concepts of deep learning, such as **forward propagation**—the propagation of data through the network, multiplying the input values by the weight of each connection for every node, and **backpropagation**—the calculation of the gradient of the loss function with respect to the weights in the matrix, and **gradient descent**—the optimization algorithm that's used to find the minimum of the loss function.

We will not delve deeply into these concepts as it isn't required for this book. However, this coverage will essentially help anyone who wants to apply deep learning to a problem.

Then, we will move on to implementing neural networks using Keras. Also, we will stick to the simplest case, which is a neural network with a single hidden layer. You will learn how to define a model in Keras, choose the **hyperparameters**—the parameters of the model that are set before training the model—and then train your model. At the end of this section, you will have the opportunity to practice what you have learned by implementing a neural network in Keras so that you can perform classification on a dataset and observe how neural networks outperform simpler models such as logistic regression.

LOGISTIC REGRESSION TO A DEEP NEURAL NETWORK

In *Chapter 1*, *Introduction to Machine Learning with Keras*, you learned about the logistic regression model, and then how to implement it as a sequential model using Keras in *Chapter 2*, *Machine Learning versus Deep Learning*. Technically speaking, logistic regression involves a very simple neural network with only one hidden layer and only one node in its hidden layer.

An overview of the logistic regression model with two-dimensional input can be seen in the following image. What you see in this image is called one **node** or **unit** in the deep learning world, which is represented by the green circle. As you may have noticed, there are some differences between logistic regression terminology and deep learning terminology. In logistic regression, we call the parameters of the model **coefficients** and **intercepts**. In deep learning models, the parameters are referred to as **weights** (w) and **biases** (b):

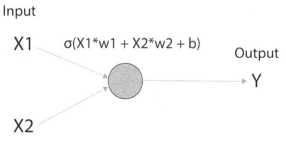

Figure 3.1: Overview of the logistic regression model with a two-dimensional input

At each node/unit, the inputs are multiplied by some weights and then a bias term is added to the sum of these weighted inputs. This can be seen in the calculation above the node in the preceding image. The **inputs** are **X1** and **X2**, the **weights** are **W1** and **W2**, and the **bias** is **b**. Next, a nonlinear function (for example, a sigmoid function in the case of a logistic regression model) is applied to the sum of the weighted inputs and the bias term is used to compute the final output of the node. In the calculation shown in the preceding image, this is **σ**. In deep learning, the nonlinear function is called the **activation function** and the output of the node is called the **activation** of that node.

It is possible to build a single-layer neural network by stacking logistic regression nodes/units on top of each other in a layer, as shown in the following image. Every value at the input layers, **X1** and **X2**, is passed to all three nodes at the hidden layer:

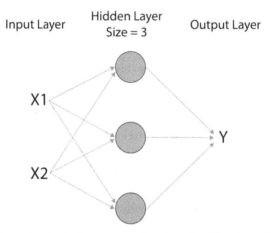

Figure 3.2: Overview of a single-layer neural network with a two-dimensional input and a hidden layer of size 3

It is also possible to build multi-layer neural networks by stacking multiple layers of processing nodes after one another, as shown in the following image. The following image shows a two-layer neural network with two-dimensional input:

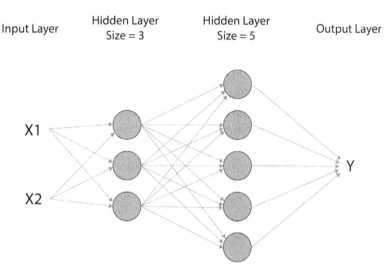

Figure 3.3: Overview of a two-layer neural network with a two-dimensional input

The preceding two images show the most common way of representing a neural network. Every neural network consists of an **input layer**, an **output layer**, and one or many **hidden layers**. If there is only one hidden layer, the network is called a **shallow neural network**. On the other hand, neural networks with many hidden layers are called **deep neural networks**, and the process of training them is called **deep learning**.

Figure 3.2 shows a neural network with only one hidden layer, so this would be a shallow neural network, whereas the neural network in *Figure 3.3* has two hidden layers, so it is a deep neural network. The input layers are generally on the left. In the case of *Figure 3.3*, these are features **X1** and **X2**, and they are input into the first hidden layer, which has three nodes. The arrows represent the weight values that are applied to the input. At the second hidden layer, the result of the first hidden layer becomes the input to the second hidden layer. The arrows between the first and second hidden layers represent the weights. The output is generally the layer on the far right and, in the case of *Figure 3.3*, is represented by the layer labeled **Y**.

> **NOTE**
>
> In some resources, you may see that a network, such as the one shown in the preceding image, is referred to as a **four-layer network**. This is because the input and output layers are counted along with the hidden layers. However, the more common convention is to count only the hidden layers, so the network we mentioned previously will be referred to as a two-layer network.

In a deep learning setting, the number of nodes in the input layer is equal to the number of features of the input data, and the number of nodes in the output layer is equal to the number of dimensions of the output data. However, you need to select the number of nodes in the hidden layers or the size of the hidden layers. If you choose a larger size layer, the model becomes more flexible and will be able to model more complex functions. This increase in flexibility comes at the cost of the need for more training data and more computations to train the model on. The parameters that are required to be selected by the developer are called **hyperparameters** and include parameters such as the number of layers and the number of nodes in each layer. Common hyperparameters to be chosen include the number of epochs to train for and the loss function to use.

In the next section, we will cover **activation functions** that are applied after each hidden layer.

ACTIVATION FUNCTIONS

In addition to the size of the layer, you need to choose an activation function for each hidden layer that you add to the model, and also do the same for the output layer. We learned about the sigmoid activation function in the logistic regression model. However, there are more options for activation functions that you can choose from when building a neural network in Keras. For example, the sigmoid activation function is a good choice as the activation function on the output layer for binary classification tasks since the result of a sigmoid function is bounded between 0 and 1. Some commonly used activation functions for deep learning are **sigmoid/logistic, tanh (hyperbolic tangent)**, and **Rectified Linear Unit (ReLU)**.

The following image shows a **sigmoid** activation function:

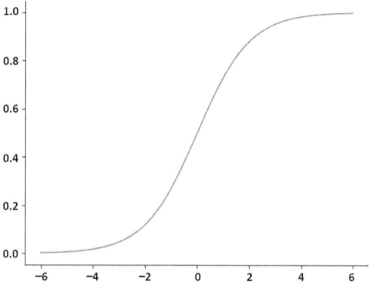

Figure 3.4: Sigmoid activation function

The following image shows a **tanh** activation function:

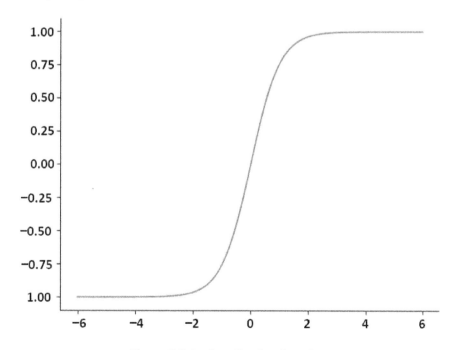

Figure 3.5: tanh activation function

The following image shows a **ReLU** activation function:

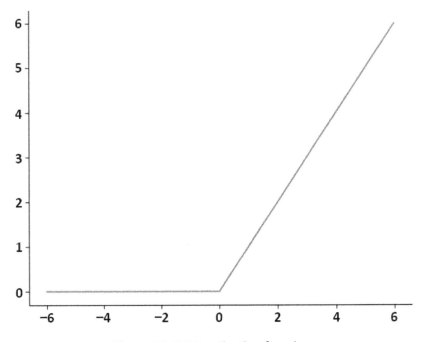

Figure 3.6: ReLU activation function

As you can see in *Figures 3.4* and *3.5*, the output of a sigmoid function is always between **0** and **1**, and the output of tanh is always between **−1** and **1**. This makes **tanh** a better choice for hidden layers since it keeps the average of the outputs in each layer close to zero. In fact, **sigmoid** is only a good choice for the **activation function** of the output layer when building a **binary classifier** since its output can be interpreted as the probability of a given input belonging to one class.

Therefore, **tanh** and **ReLU** are the most common choices of activation function for hidden layers. It turns out that the learning process is faster when using the **ReLU activation function** because it has a fixed derivative (or slope) for an input greater than **0**, and a slope of **0** everywhere else.

> **NOTE**
>
> You can read more about all the available choices for activation functions in Keras here: https://keras.io/activations/.

FORWARD PROPAGATION FOR MAKING PREDICTIONS

Neural networks make a prediction about the output by performing **forward propagation**. Forward propagation entails the computations that are performed on the input in every layer of a neural network until the output layer is reached. It is best to understand forward propagation through an example.

Let's go through forward propagation equations one by one for a two-layer neural network (shown in the following image) where the input data is two-dimensional, and the output data is a one-dimensional binary class label. The activation functions for layer 1 and layer 2 will be tanh, and the activation function in the output layer is sigmoid.

The following image shows the weights and biases for each layer as matrices and vectors with proper indexes. For each layer, the number of rows in the weight's matrix is equal to the number of nodes in the previous layer, and the number of columns is equal to the number of nodes in that layer.

For example, **W1** has two rows and three columns because the input to layer 1 is the input layer, **X**, which has two columns, and layer 1 has three nodes. Likewise, **W2** has three rows and three columns because the input to layer 2 is layer 1, which has two nodes, and layer 2 has five nodes. The bias, however, is always a vector with a size equal to the number of nodes in that layer. The total number of parameters in a deep learning model is equal to the total number of elements in all the weights' matrices and the biases' vectors:

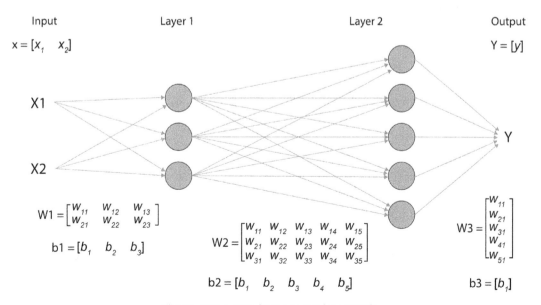

Figure 3.7: A two-layer neural network

An example of performing all the steps for forward propagation according to the neural network outlined in the preceding image is as follows.

Steps to perform forward propagation:

1. **X** is the network input to the network in the preceding image, so it is the input for the first hidden layer. First, the input matrix, **X**, is the matrix multiplied by the weight matrix for layer 1, **W1**, and the bias, **b1**, is added:

    $z1 = X*W1 + b1$

2. Next, the layer 1 output is computed by applying an activation function to $z1$, which is the output of the previous step:

    $a1 = tanh(z1)$

3. **a1** is the output of layer 1 and is called the **activation of layer 1**. The output of layer 1 is, in fact, the **input** for layer 2. Next, the activation of layer 1 is the matrix multiplied by the weight matrix for layer 2, **W2**, and the bias, **b2**, is added:

    $z2 = a1 * W2 + b2$

4. The layer 2 output/activation is computed by applying an activation function to **z2**:

    $a2 = tanh(z2)$

5. The output of layer 2 is, in fact, the input for the next layer (the network output layer here). Following this, the activation of layer 2 is the matrix multiplied by the weight matrix for the output layer, **W3**, and the bias, **b3**, is added:

    $z3 = a2 * W3 + b3$

6. Finally, the network output, Y, is computed by applying the sigmoid activation function to **z3**:

    $Y = sigmoid(z3)$

The total number of parameters in this model is equal to the sum of the number of elements in **W1**, **W2**, **W3**, **b1**, **b2**, and **b3**. Therefore, the number of parameters can be calculated by summing the parameters in each of the parameters in weight matrices and biases, which is equal to 6 + 15 + 5 + 3 + 5 + 1 = 35. These are the parameters that need to be learned in the process of deep learning.

Now that we have learned about the forward propagation step, we have to evaluate our model and compare it to the real target values. To achieve that, we will use a loss function, which we will cover in the next section. Here, we will learn about some common loss functions that we can use for classification and regression tasks.

LOSS FUNCTION

When learning the optimal parameters (weights and biases) of a model, we need to define a function to measure error. This function is called the **loss function** and it provides us with a measure of how different network-predicted outputs are from the real outputs in the dataset.

The loss function can be defined in several different ways, depending on the problem and the goal. For example, in the case of a classification problem, one common way to define loss is to compute the proportion of misclassified inputs in the dataset and use that as the probability of the model making an error. On the other hand, in the case of a regression problem, the loss function is usually defined by computing the distance between the predicted outputs and their corresponding real outputs, and then averaging over all the examples in the dataset.

Brief descriptions of some commonly used loss functions that are available in Keras are as follows:

- **mean_squared_error** is a loss function for regression problems that calculates `(real output - predicted output)^2` for each example in the dataset and then returns their average.

- **mean_absolute_error** is a loss function for regression problems that calculates `abs (real output - predicted output)` for each example in the dataset and then returns their average.

- **mean_absolute_percentage_error** is a loss function for regression problems that calculates `abs [(real output - predicted output) / real output]` for each example in the dataset and then returns their average, multiplied by 100%.

- **binary_crossentropy** is a loss function for two-class/binary classification problems. In general, the cross-entropy loss is used for calculating the loss for models where the output is a probability number between **0** and **1**.

- **categorical_crossentropy** is a loss function for multi-class (more than two classes) classification problems.

> **NOTE**
>
> You can read more about all the available choices for loss functions in Keras here: https://keras.io/losses/.

During the training process, we keep changing the model parameters until the minimum difference between the model-predicted outputs and the real outputs is reached. This is called an **optimization process**, and we will learn more about how it works in later sections. For neural networks, we use backpropagation to compute the derivatives of the loss function with respect to the weights.

BACKPROPAGATION FOR COMPUTING DERIVATIVES OF LOSS FUNCTION

Backpropagation is the process of performing the chain rule of calculus from the output layer to the input layer of a neural network in order to compute the derivatives of the loss function with respect to the model parameters in each layer. The derivative of a function is simply the slope of that function. We are interested in the slope of the loss function because it provides us with the direction in which model parameters need to change in order for the loss value to be minimized.

The chain rule of calculus states that if, for example, **z** is a function of **y**, and **y** is a function of **x**, then the derivative of **z** with respect to **x** can be reached by multiplying the derivative of **z** with respect to **y** by the derivative of **y** with respect to **x**. This can be written as follows:

$dz/dx = dz/dy * dy/dx$

In deep neural networks, the loss function is a function of predicted outputs. We can show this through the equation given here:

$loss = L(y\_predicted)$

On the other hand, according to forward propagation equations, the output predicted by the model is a function of the model parameters—that is, the weights and biases in each layer. Therefore, according to the chain rule of calculus, the derivative of the loss with respect to the model parameters can be computed by multiplying the derivative of the loss with respect to the predicted output by the derivative of the predicted output with respect to the model parameters.

In the next section, we will learn how the optimal weight parameters are modified when given the derivatives of the loss function with respect to the weights.

GRADIENT DESCENT FOR LEARNING PARAMETERS

In this section, we will learn how a deep learning model learns its optimal parameters. Our goal is to update the weight parameters so that the loss function is minimized. This will be an iterative process in which we continue to update the weight parameters so that the loss function is at a minimum. This process is called **learning parameters** and it is done through the use of an optimization algorithm. One very common optimization algorithm that's used for learning parameters in machine learning is **gradient descent**. Let's see how gradient descent works.

If we plot the average of loss over all the examples in the dataset for all the possible values of the model parameters, it is usually a convex shape (such as the one shown in the following plot). In gradient descent, our goal is to find the minimum point (**Pt**) on the plot. The algorithm starts by initializing the model parameters with some random values (**P1**). Then, it computes the loss and the derivatives of the loss with respect to the parameters at that point. As we mentioned previously, the derivative of a function is, in fact, the slope of the function. After computing the slope at an initial point, we have the direction in which we need to update the parameters.

The hyperparameter, called the **learning rate** (**alpha**), determines how big a step the algorithm will take from the initial point. After selecting the proper alpha value, the algorithm updates the parameters from their initial values to the new values (shown as point **P2** in the following plot). As shown in the following plot, **P2** is closer to the target point, and if we keep moving in that direction, we will eventually get to the target point, **Pt**. The algorithm computes the slope of the function again at **P2** and takes another step.

This process is repeated until the slope is equal to zero and therefore no direction for further movement is provided:

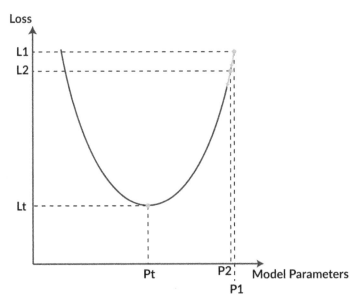

Figure 3.8: A schematic view of the gradient descent algorithm finding the set of parameters that minimize loss

The pseudocode for the gradient descent algorithm is provided here:

```
Initialize all the weights (w) and biases (b) arbitrarily
Repeat Until converge {
Compute loss given w and b
Compute derivatives of loss with respect to w (dw), and with respect to b
(db) using backpropagation
Update w to w - alpha * dw
Update b to b - alpha * db
}
```

To summarize, the following steps are repeated when training a deep neural network (after initializing the parameters to some random values):

1. Use forward propagation and the current parameters to predict the outputs for the entire dataset.

2. Use the predicted outputs to compute the loss over all the examples.

3. Use backpropagation to compute the derivatives of the loss with respect to the weights and biases at each layer.

4. Update the weights and biases using the derivative values and the learning rate.

What we discussed here was the standard gradient descent algorithm, which computes the loss and the derivatives using the entire dataset in order to update the parameters. There is another version of gradient descent called **stochastic gradient descent** (**SGD**), which computes the loss and the derivatives each time using a subset or a batch of data examples only; therefore, its learning process is faster than standard gradient descent.

> **NOTE**
>
> Another common choice is an optimization algorithm called **Adam**. Adam usually outperforms **SGD** when training deep learning models. As we've already learned, **SGD** uses a single hyperparameter (called a **learning rate**) to update the parameters. However, Adam improves this process by using a learning rate, a weighted average of gradients, and a weighted average of squared gradients to update the parameters at each iteration.

Usually, when building a neural network, you need to choose two hyperparameters (called the **batch size** and the number of **epochs**) for your optimization process. The `batch_size` argument determines the number of data examples to be included at each iteration of the optimization algorithm. `batch_size=None` is equivalent to the standard version of gradient descent, which uses the entire dataset in each iteration. The **epochs** argument determines how many times the optimization algorithm passes through the entire training dataset before it stops.

For example, imagine we have a dataset of size **n=400**, and we choose `batch_size=5` and **epochs=20**. In this case, the optimizer will have **400/5 = 80** iterations in one pass through the entire dataset. Since it is supposed to go through the entire dataset **20** times, it will have **80 * 20** iterations in total.

> **NOTE**
>
> When building a model in Keras, you need to choose the type of optimizer to be used when training your model. There are some other options other than SGD and Adam available in Keras. You can read more about all the possible options for optimizers in Keras here: https://keras.io/optimizers/.

> **NOTE**
>
> All the activities and exercises in this chapter will be developed in a Jupyter notebook. Please download this book's GitHub repository, along with all the prepared templates, from https://packt.live/39pOUMT.

EXERCISE 3.01: NEURAL NETWORK IMPLEMENTATION WITH KERAS

In this exercise, you will learn the step-by-step process of implementing a neural network using Keras. Our simulated dataset represents various measurements of trees, such as height, the number of branches, the girth of the trunk at the base, and more, that are found in a forest. Our goal is to classify the records into either deciduous or coniferous type trees based on the measurements given. First, execute the following code block to load a simulated dataset of **10000** records that consist of two classes, representing the two tree species, where each data example has **10** feature values:

```
import numpy as np
import pandas as pd
X = pd.read_csv('../data/tree_class_feats.csv')
y = pd.read_csv('../data/tree_class_target.csv')

# Print the sizes of the dataset
print("Number of Examples in the Dataset = ", X.shape[0])
print("Number of Features for each example = ", X.shape[1])
print("Possible Output Classes = ", np.unique(y))
```

Expected output:

```
Number of Examples in the Dataset = 10000
Number of Features for each example = 10
Possible Output Classes = [0 1]
```

Since each data example in this dataset can only belong to one of the two classes, this is a binary classification problem. Binary classification problems are very important and very common in real-life scenarios. For example, let's assume that the examples in this dataset represent the measurement results for **10000** trees from a forest. The goal is to build a model using this dataset to predict whether the species of each tree that's measured is a deciduous or coniferous species of tree. The **10** features for the trees can include predictors such as height, number of branches, and girth of the trunk at the base.

The output class **0** means that the tree is a coniferous species of tree, while the output class **1** means that the tree is a deciduous species of tree.

Now, let's go through the steps for building and training a Keras model to perform the classification:

1. Set a seed in **numpy** and **tensorflow** and define your model as a Keras sequential model. **Sequential** models are, in fact, stacks of layers. After defining the model, we can add as many layers to it as desired:

```
from keras.models import Sequential
from tensorflow import random
np.random.seed(42)
random.set_seed(42)
model = Sequential()
```

2. Add one hidden layer of size **10** with an activation function of type **tanh** to your model (remember that the input dimension is equal to **10**). There are different types of layers available in Keras. For now, we will use only the simplest type of layer, called the **Dense** layer. A Dense layer is equivalent to the **fully connected layers** that we have seen in all the examples so far:

```
from keras.layers import Dense, Activation
model.add(Dense(10, activation='tanh', input_dim=10))
```

3. Add another hidden layer, this time of size **5** and with an activation function of type **tanh**, to your model. Please note that the input dimension argument is only provided for the first layer since the input dimension for the next layers is known:

```
model.add(Dense(5, activation='tanh'))
```

4. Add the output layer with the **sigmoid** activation function. Please note that the number of units in the output layer is equal to the output dimension:

```
model.add(Dense(1, activation='sigmoid'))
```

5. Ensure that the loss function is binary cross-entropy and that the optimizer is **SGD** for training the model using the **compile()** method and print out a summary of the model to see its architecture:

```
model.compile(optimizer='sgd', loss='binary_crossentropy', \
              metrics=['accuracy'])
model.summary()
```

The following image shows the output of the preceding code:

```
Model: "sequential_1"

Layer (type)                    Output Shape                Param #
=====================================================================
dense_1 (Dense)                 (None, 10)                  110

dense_2 (Dense)                 (None, 5)                   55

dense_3 (Dense)                 (None, 1)                   6
=====================================================================
Total params: 171
Trainable params: 171
Non-trainable params: 0
```

Figure 3.9: A summary of the model that was created

6. Train your model for **100** epochs and set a **batch_size** equal to 5 and a
   validation_split equal to **0.2**, and then set **shuffle** equal to **false**
   using the **fit()** method. Remember that you need to pass the input data, **X**,
   and its corresponding outputs, **y**, to the **fit()** method to train the model. Also,
   keep in mind that training a network may take a long time, depending on the size
   of the dataset, the size of the network, the number of epochs, and the number of
   CPUs or GPUs available. Save the results to a variable named **history**:

```
history = model.fit(X, y, epochs=100, batch_size=5, \
                    verbose=1, validation_split=0.2, \
                    shuffle=False)
```

The **verbose** argument can take any of these three values: **0**, **1**, or **2**.
By choosing **verbose=0**, no information will be printed during training.
verbose=1 will print a full progress bar at every iteration, while **verbose=2**
will print only the epoch number:

```
Epoch 96/100
8000/8000 [==============================] - 2s 227us/step - loss: 0.1372 - accuracy: 0.9469 - val_loss: 0.1488 - val
_accuracy: 0.9405
Epoch 97/100
8000/8000 [==============================] - 2s 220us/step - loss: 0.1397 - accuracy: 0.9481 - val_loss: 0.1457 - val
_accuracy: 0.9385
Epoch 98/100
8000/8000 [==============================] - 2s 258us/step - loss: 0.1381 - accuracy: 0.9473 - val_loss: 0.1495 - val
_accuracy: 0.9405
Epoch 99/100
8000/8000 [==============================] - 2s 217us/step - loss: 0.1377 - accuracy: 0.9484 - val_loss: 0.1503 - val
_accuracy: 0.9385
Epoch 100/100
8000/8000 [==============================] - 2s 236us/step - loss: 0.1386 - accuracy: 0.9467 - val_loss: 0.1422 - val
_accuracy: 0.9440
```

Figure 3.10: The loss details of the last 5 epochs out of 400

7. Print the accuracy and loss of the model on the training and validation data as a function of the epoch:

```
import matplotlib.pyplot as plt
%matplotlib inline

# Plot training & validation accuracy values
plt.plot(history.history['accuracy'])
plt.plot(history.history['val_accuracy'])
plt.title('Model accuracy')
plt.ylabel('Accuracy')
plt.xlabel('Epoch')
plt.legend(['Train', 'Validation'], loc='upper left')
plt.show()

# Plot training & validation loss values
plt.plot(history.history['loss'])
plt.plot(history.history['val_loss'])
plt.title('Model loss')
plt.ylabel('Loss')
plt.xlabel('Epoch')
plt.legend(['Train', 'Validation'], loc='upper left')
plt.show()
```

The following image shows the output of the preceding code:

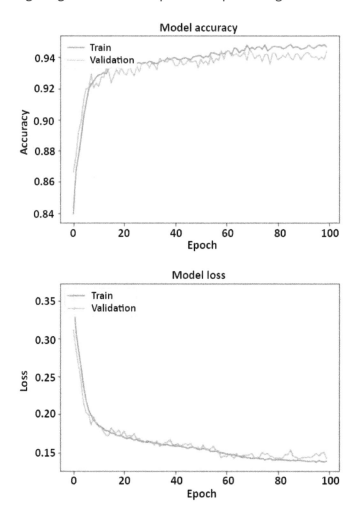

Figure 3.11: The model's accuracy and loss as a function of an epoch during the training process

8. Use your trained model to predict the output class for the first 10 input data examples (**X.iloc[0:10,:]**):

```
y_predicted = model.predict(X.iloc[0:10,:])
```

You can print the predicted classes using the following code block:

```
# print the predicted classes
print("Predicted probability for each of the "\
    "examples belonging to class 1: "),
```

```
print(y_predicted)
print("Predicted class label for each of the examples: "),
print(np.round(y_predicted))
```

Expected output:

```
Predicted probability for each of the examples belonging to class 1:
[[0.00354007]
 [0.8302744 ]
 [0.00316998]
 [0.95335543]
 [0.99479216]
 [0.00334176]
 [0.43222323]
 [0.00391936]
 [0.00332899]
 [0.99759173]
Predicted class label for each of the examples:
[[0.]
 [1.]
 [0.]
 [1.]
 [1.]
 [0.]
 [0.]
 [0.]
 [0.]
 [1.]]
```

Here, we used the trained model to predict the output for the first 10 tree species in the dataset. As you can see, the model predicted that the second, fourth, fifth, and tenth trees were predicted as the species of class 1, which is deciduous.

NOTE

To access the source code for this specific section, please refer to https://packt.live/2YX3fxX.

You can also run this example online at https://packt.live/38pztVR.

Please note that you can extend these steps by adding more hidden layers to your network. In fact, you can add as many layers as you want to your model before adding the output layer. However, the input dimension argument is only provided for the first layer since the input dimension for the next layers is known. Now that you have learned how to implement a neural network in Keras, you are ready to practice with them further by implementing a neural network that can perform classification in the following activity.

ACTIVITY 3.01: BUILDING A SINGLE-LAYER NEURAL NETWORK FOR PERFORMING BINARY CLASSIFICATION

In this activity, we will use a Keras sequential model to build a binary classifier. The simulated dataset provided represents the testing results of the production of aircraft propellers. Our target variable will be the results of the manual inspection of the propellers, designated as either "pass" (represented as a value of 1) or "fail" (represented as a value of 0).

Our goal is to classify the testing results into either "pass" or "fail" classes to match the manual inspections. We will use models with different architectures and observe the visualization of the different models' performance. This will help you gain a better sense of how going from one processing unit to a layer of processing units changes the flexibility and performance of the model.

Assume that this dataset contains two features representing the test results of two different tests inspecting the aircraft propellers of over **3000** propellers (the two features are normalized to have a mean of zero). The output is the likelihood of the propeller passing the test, with 1 representing a pass and zero representing a fail. The company would like to rely less on time-consuming, error-prone manual inspections of the aircraft propellers and shift resources to developing automated tests to assess the propellers faster. Therefore, the goal is to build a model that can predict whether an aircraft propeller will pass the manual inspection when given the results from the two tests. In this activity, you will first build a logistic regression model, then a single-layer neural network with three units, and finally a single-layer neural network with six units, to perform the classification. Follow these steps to complete this activity:

1. Import the required packages:

```
# import required packages from Keras
from keras.models import Sequential
from keras.layers import Dense, Activation
import numpy as np
import pandas as pd
```

```
from tensorflow import random
from sklearn.model_selection import train_test_split
# import required packages for plotting
import matplotlib.pyplot as plt
import matplotlib
%matplotlib inline
import matplotlib.patches as mpatches
# import the function for plotting decision boundary
from utils import plot_decision_boundary
```

> **NOTE**
>
> You will need to download the **utils.py** file from the GitHub repository and save it into your activity folder in order for the utils import statement to work correctly. You can find the file here: https://packt.live/31EumPY.

2. Set up a seed for a random number generator so that the results will be reproducible:

```
"""
define a seed for random number generator so the result will be
reproducible
"""

seed = 1
```

> **NOTE**
>
> The triple-quotes (**"""**) shown in the code snippet above are used to denote the start and end points of a multi-line code comment. Comments are added into code to help explain specific bits of logic.

3. Load the dataset using the **read_csv** function from the **pandas** library. Print the **X** and **Y** sizes and the number of examples in the training dataset using **feats.shape**, **target.shape**, and **feats.shape[0]**:

```
feats = pd.read_csv('outlier_feats.csv')
target = pd.read_csv('outlier_target.csv')
print("X size = ", feats.shape)
print("Y size = ", target.shape)
print("Number of examples = ", feats.shape[0])
```

4. Plot the dataset using the following code:

```
plt.scatter(feats[:,0], feats[:,1], \
            s=40, c=Y, cmap=plt.cm.Spectral)
```

5. Implement a logistic regression model as a sequential model in Keras. Remember that the activation function for binary classification needs to be sigmoid.

6. Train the model with **optimizer='sgd'**, **loss='binary_crossentropy'**, **batch_size = 5**, **epochs = 100**, and **shuffle=False**. Observe the loss values in each iteration by using **verbose=1** and **validation_split=0.2**.

7. Plot the decision boundary of the trained model using the following code:

```
plot_decision_boundary(lambda x: model.predict(x), \
                       X_train, y_train)
```

8. Implement a single-layer neural network with three nodes in the hidden layer and the **ReLU activation function** for **200** epochs. It is important to remember that the activation function for the output layer still needs to be sigmoid since it is a binary classification problem. Choosing **ReLU** or having no activation function for the output layer will not produce outputs that can be interpreted as class labels. Train the model with **verbose=1** and observe the loss in every iteration. After the model has been trained, plot the decision boundary and evaluate the loss and accuracy on the test dataset.

9. Repeat *step 8* for the hidden layer of **size 6** and **400** epochs and compare the final loss value and the decision boundary plot.

10. Repeat *steps 8* and *9* using the **tanh** activation function for the hidden layer and compare the results with the models with **relu** activation. Which activation function do you think is a better choice for this problem?

> **NOTE**
>
> The solution for this activity can be found on page 362.

In this activity, you observed how stacking multiple processing units in a layer can create a much more powerful model than a single processing unit. This is the basic reason why neural networks are such powerful models. You also observed that increasing the number of units in the layer increases the flexibility of the model, meaning a non-linear separating decision boundary can be estimated more precisely.

However, a model with more processing units takes longer to learn the patterns, requires more epochs to be trained, and can overfit the training data. As such, neural networks are computationally expensive models. You also observed that using the tanh activation function results in a slower training process in comparison to using the **ReLU activation function**.

In this section, we created various models and trained them on our data. We observed that some models performed better than others by evaluating them on the data that they were trained on. In the next section, we learn about some alternative methods we can use to evaluate our models that provide an unbiased evaluation.

MODEL EVALUATION

In this section, we will move on to multi-layer or deep neural networks while learning about techniques for assessing the performance of a model. As you may have already realized, there are many hyperparameter choices to be made when building a deep neural network.

Some of the challenges of applied deep learning include how to find the right values for the number of hidden layers, the number of units in each hidden layer, the type of activation function to use for each layer, and the type of optimizer and loss function for training the network. Model evaluation is required when making these decisions. By performing model evaluation, you can say whether a specific deep architecture or a specific set of hyperparameters is working poorly or well on a particular dataset, and therefore decide whether to change them or not.

Furthermore, you will learn about **overfitting** and **underfitting**. These are two very important issues that can arise when building and training deep neural networks. Understanding the concepts of overfitting and underfitting and whether they are happening in practice is essential when it comes to finding the right deep neural network for a particular problem and improving its performance as much as possible.

EVALUATING A TRAINED MODEL WITH KERAS

In the previous activity, we plotted the decision boundary of the model by predicting the output for every possible value of the input. Such visualization of model performance was possible because we were dealing with two-dimensional input data. The number of features or measurements in the input space is almost always way more than two, and so visualization by 2D plotting is not an option. One way to figure out how well a model is doing on a particular dataset is to compute the overall loss when predicting outputs for many examples. This can be done by using the **evaluate()** method in Keras, which receives a set of inputs (**X**) and their corresponding outputs (**y**), and calculates and returns the overall loss of the model on the inputs, **X**.

For example, let's consider a case of building a neural network with two hidden layers of sizes **8** and **4**, respectively, in order to perform binary or two-class classification. The available data points and their corresponding class labels are stored in **X**, **y** arrays. We can build and train the mentioned model as follows:

```
model = Sequential()
model.add(Dense(8, activation='tanh', input_dim=2))
model.add(Dense(4, activation='tanh'))
model.add(Dense(1, activation='sigmoid'))
model.compile(optimizer='sgd', loss='binary_crossentropy')
model.fit(X, y, epochs=epochs, batch_size=batch_size)
```

Now, instead of using **model.predict()** to predict the output for a given set of inputs, we can evaluate the overall performance of the model by calculating the loss on the whole dataset by writing the following:

```
model.evaluate(X, y, batch_size=None, verbose=0)
```

If you include other metrics, such as accuracy, when defining the **compile()** method for the model, the **evaluate()** method will return those metrics along with the loss when it is called. For example, if we add metrics to the **compile()** arguments, as shown in the following code, then calling the **evaluate()** method will return the overall loss and the overall accuracy of the trained model on the whole dataset:

```
model.compile(optimizer='sgd', loss='binary_crossentropy', \
              metrics=['accuracy'])
model.evaluate(X, y, batch_size=None, verbose=0)
```

> **NOTE**
>
> You can check out all the possible options for the **metrics** argument in Keras here: https://keras.io/metrics/.

In the next section, we will learn about splitting the dataset into training and test datasets. Much like we did in *Chapter 1, Introduction to Machine Learning with Keras*, training on separate data for evaluation can provide an unbiased evaluation of your model's performance.

SPLITTING DATA INTO TRAINING AND TEST SETS

In general, evaluating a model on the same dataset that has been used for training the model is a methodological mistake. Since the model has been trained to reduce the errors on this dataset, performing an evaluation on it will result in a biased estimation of the model performance. In other words, the error rate on the dataset that has been used for training is always an underestimation of the error rate on new unseen examples.

On the other hand, when building a machine learning model, the goal is not to achieve good performance on the training data only, but to achieve good performance on future examples that the model has not seen during training. That is why we are interested in evaluating the performance of a model using a dataset that has not been used for training the model.

One way to achieve this is to split the available dataset into two sets: a training set and a test set. The training set is used to train the model, while the test set is used for performance evaluation. More precisely, the role of the training set is to provide enough examples for the model that it will learn the relations and patterns in the data, while the role of the test set is to provide us with an unbiased estimation of the model performance on new unseen examples. The common practice in machine learning is to perform **70%-30%** or **80%-20%** splitting for training-test sets. This is usually the case for relatively small datasets. When dealing with a dataset with millions of examples in which the goal is to train a large deep neural network, the training-test splitting can be done using **98%-2%** or **99%-1%** ratios.

The following image shows the division of a dataset into a training set and test set. Notice that there is no overlap between the **training** set and **test** set:

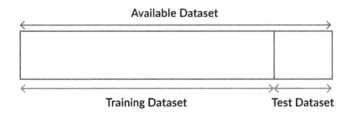

Figure 3.12: Illustration of splitting a dataset into training and test sets

You can easily perform splitting on your dataset using scikit-learn's **train_test_split** function. For example, the following code will perform a **70%-30%** training-test split on the dataset:

```
from sklearn.model_selection import train_test_split
X_train, X_test, \
y_train, y_test = train_test_split(X, y, test_size=0.3, \
                                   random_state=None)
```

The **test_size** argument represents the proportion of the dataset to be kept in the test set, so it should be between **0** and **1**. By assigning an **int** to the **random_state** argument, you can choose the seed to be used to generate the random split between the training and test sets.

After splitting the data into training and test sets, we can change the code from the previous section by providing only the training set as an argument to **fit()**:

```
model = Sequential()
model.add(Dense(8, activation='tanh', input_dim=2))
model.add(Dense(4, activation='tanh'))
model.add(Dense(1, activation='sigmoid'))
model.compile(optimizer='sgd', \
              loss='binary_crossentropy')
model.fit(X_train, y_train, epochs=epochs, \
          batch_size=batch_size)
```

Now, we can compute the model error rate on the training set and the test set separately:

```
model.evaluate(X_train, y_train, batch_size=None, \
               verbose=0)
model.evaluate(X_test, y_test, batch_size=None, \
               verbose=0)
```

Another way of doing the splitting is by including the **validation_split** argument for the **fit()** method in Keras. For example, by only changing the **model.fit(X, y)** line in the code from the previous section to **model.fit(X, y, validation_split=0.3)**, the model will keep the last 30% of the data examples in a separate test set. It will only train the model on the other 70% of the samples, and it will evaluate the model on the training set and the test set at the end of each epoch. In doing so, it would be possible to observe the changes in the training error rate, as well as the test error rate as the training progresses.

The reason that we want to have an unbiased evaluation of our model is so that we can see where there is room for improvement. Since neural networks have so many parameters to learn and can learn complex functions, they can often overfit to the training data and learn the noise in the training data, which can prevent the model from performing well on new, unseen data. The next section will explore these concepts in detail.

UNDERFITTING AND OVERFITTING

In this section, you will learn about two issues you may face when building a machine learning model that needs to fit into a dataset. These issues are called **overfitting** and **underfitting** and are similar to the concepts of **bias** and **variance** for a model.

In general, if a model is not flexible enough to learn the relations and patterns in a dataset, there will be a high training error. We can call such a model a model with high bias. On the other hand, if a model is too flexible for a given dataset, it will learn the noise in the training data, as well as the relations and patterns in the data. Such a system will cause a large increase in the test error in comparison to the training error. We mentioned previously that it is always the case that the test error is slightly higher than the training error.

However, having a large gap between the test error and the training error is an indicator of a system with high variance. In data analysis, neither of these situations (**high bias** and **high variance**) are desirable. In fact, the aim is to find the model with the lowest possible amount of bias and variance at the same time.

For example, let's consider a dataset that represents the normalized locations of the sightings of two species of butterfly, as shown in the following plot. The goal is to find a model that can separate these two species of the butterfly when given the location of their sighting. Clearly, the separating line between the two classes is not linear. Therefore, if we choose a simple model such as logistic regression (a neural network with one hidden layer of size one) to perform the classification on this dataset, we will get a linear separating line/decision boundary between two classes that is unable to capture the true pattern in the dataset:

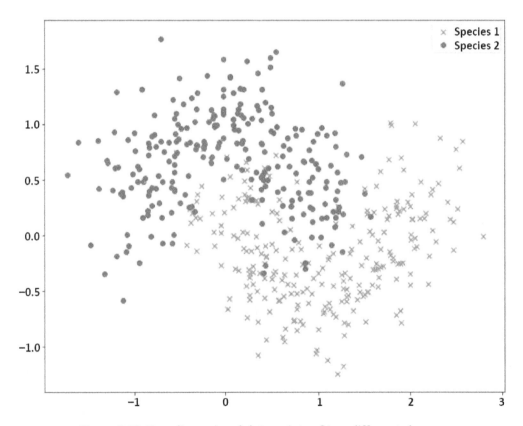

Figure 3.13: Two-dimensional data points of two different classes

The following plot illustrates the decision boundary that's achieved by such a model. By evaluating this model, it will be observed that the training error rate is high and that the test error rate is slightly higher than the training error. Having a high training error rate is indicative of a model with high bias while having a slight difference between the training error and test error is representative of a low-variance model. This is a clear case of underfitting; the model fails to fit the true separating line between the two classes:

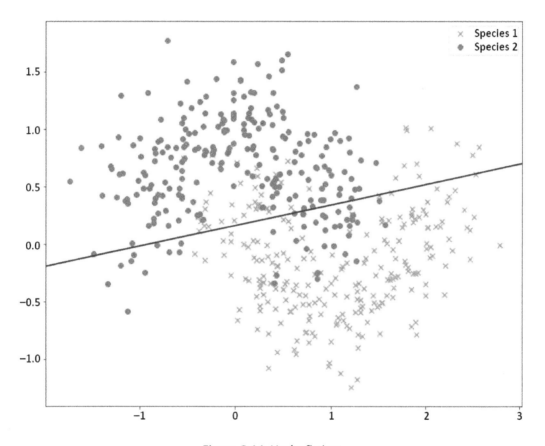

Figure 3.14: Underfitting

If we increase the flexibility of the neural network by adding more layers to it and increase the number of units in each layer, we can train a better model and succeed in capturing the non-linearity in the decision boundary. Such a model can be seen in the following plot. This is a model with a low training error rate and low-test error rate (again, the test error rate is slightly higher than the training error rate). Having a low training error rate and a slight difference between the test error rate and the training error rate is indicative of a model with low bias and low variance. A model with low bias and low variance represents the right amount of fitting for a given dataset:

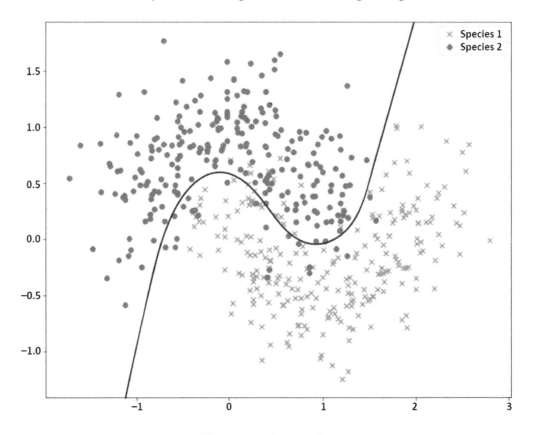

Figure 3.15: Correct fit

But what will happen if we increase the flexibility of the neural network even more? By adding too much flexibility to the model, it will learn not only the patterns and relations in the training data but also the noise in them. In other words, the model will fit each individual training example as opposed to fitting only to the overall trends and relations in them. The following plot shows such a system. Evaluating this model will show a very low training error rate and a high-test error rate (with a large difference between the training error rate and test error rate). This is a model with low bias and high variance, and this situation is called overfitting:

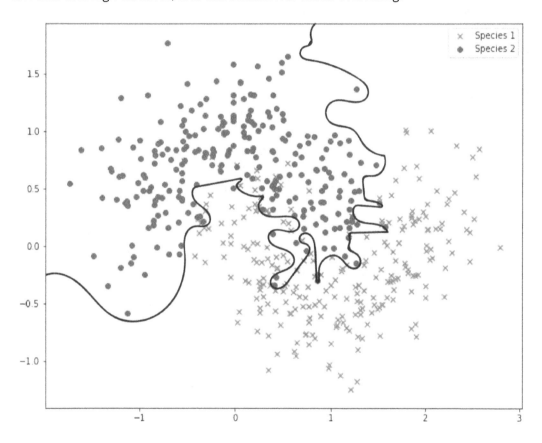

Figure 3.16: Overfitting

Evaluating the model on both the training set and the test set and comparing their error rates provide valuable information on whether the current model is right for a given dataset. Also, in cases where the current model is not fitting the dataset correctly, it is possible to determine whether it is overfitting or underfitting to the data and change the model accordingly to find the right model. For example, if the model is underfitting, you can make the network larger. On the other hand, if the model is overfitting, you can reduce the overfitting by making the network smaller or providing more training data to it. There are many methods that can be implemented to prevent underfitting or overfitting in practice, one of which we will explore in the next section.

EARLY STOPPING

Sometimes, the flexibility of a model is right for the dataset but overfitting or underfitting is still happening. This is because we are training the model for either too many iterations or too few iterations. When using an iterative optimizer such as `gradient descent`, the optimizer tries to fit the training data better and better in every iteration. Therefore, if we keep updating the parameters after the patterns in the data are learned, it will start fitting to the individual data examples.

By observing the training and test error rates in every iteration, it is possible to determine when the network is starting to overfit to the training data and stop the training process before this happens. Regions associated with `underfitting` and `overfitting` have been labeled on the following plot. The correct number of iterations for training the model can be determined from the region at which the test error rate has its lowest value. We labeled this region as the right fit on the plot and it can be seen that, in this region, both the `training error rate` and the `test error rate` are low:

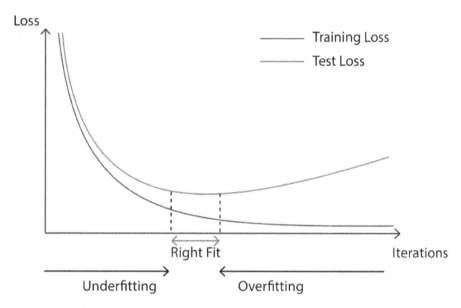

Figure 3.17: Plot of training error rate and test error rate while training a model

You can easily store the values for training loss and test loss in every epoch while training with Keras. To do this, you need to provide the test set as the **validation_data** argument when defining the **fit()** method for the model and store it in a **history** dictionary:

```
history = model.fit(X_train, y_train, validation_data=(X_test, y_test))
```

You can plot the values stored in **history** later to find the correct number of iterations to train your model with:

```
import matplotlib.pyplot as plt
import matplotlib


# plot training loss
plt.plot(history.history['loss'])
# plot test loss
plt.plot(history.history['val_loss'])
```

In general, since deep neural networks are highly flexible models, the chance of overfitting happening is very high. There is a whole group of techniques, called **regularization** techniques, that have been developed to reduce overfitting in machine learning models in general, and deep neural networks in particular. You will learn more about these techniques in *Chapter 5, Improving Model Accuracy*. In the next activity, we will put our understanding into practice and attempt to find the optimal number of epochs to train for so as to prevent overfitting.

ACTIVITY 3.02: ADVANCED FIBROSIS DIAGNOSIS WITH NEURAL NETWORKS

In this activity, you are going to use a real dataset to predict whether a patient has advanced fibrosis based on measurements such as age, gender, and BMI. The dataset consists of information for 1,385 patients who underwent treatment dosages for hepatitis C. For each patient, **28** different attributes are available, as well as a class label, which can only take two values: **1**, indicating advanced fibrosis, and **0**, indicating no indication of advanced fibrosis. This is a binary/two-class classification problem with an input dimension equal to **28**.

In this activity, you will implement different deep neural network architectures to perform this classification. Plot the trends in the training error rates and test error rates and determine how many epochs the final classifier needs to be trained for:

> **NOTE**
>
> The dataset that's being used in this activity can be found here:
> https://packt.live/39pOUMT.

Figure 3.18: Schematic view of the binary classifier for a diabetes diagnosis

Follow these steps to complete this activity:

1. Import all the necessary dependencies. Load the dataset from the **data** subfolder of the **Chapter03** folder from GitHub:

```
X = pd.read_csv('../data/HCV_feats.csv')
y = pd.read_csv('../data/HCV_target.csv')
```

2. Print the number of examples in the dataset, the number of features available, and the possible values for the class labels.

3. Scale the data using the **StandardScalar** function from **sklearn. preprocessing** and split the dataset into the training set and test set with an **80:20** ratio. Then, print the number of examples in each set after splitting.

4. Implement a shallow neural network with one hidden layer of size 3 and a **tanh** activation function to perform the classification. Compile the model with the following values for the hyperparameters: **optimizer = 'sgd', loss = 'binary_crossentropy', metrics = ['accuracy']**

5. Fit the model with the following hyperparameters and store the values for training error rate and test error rate during the training process: **batch_size = 20, epochs = 100, validation_split=0.1**, and **shuffle=False**.

6. Plot the training error rate and test error rate for every epoch of training. Use the plot to determine at which epoch the network is starting to overfit to the dataset. Also, print the values of the best accuracy that were reached on the training set and on the test set, as well as the loss and accuracy that were evaluated on the test dataset.

7. Repeat *steps 4* and *5* for a deep neural network with two hidden layers (the first layer of size 4 and the second layer of size 3) and a **'tanh'** activation function for both layers in order to perform the classification.

> **NOTE**
>
> The solution for this activity can be found on page 374.

Please note that both models were able to achieve better **accuracy** on the **training** or **validation** set compared to the **test** set, and the **training error rate** kept decreasing when it was trained for a significant number of epochs. However, the **validation error rate** decreased during training to a certain value, and after that, it started increasing, which is indicative of **overfitting** to the **training** data. The maximum validation accuracy corresponds to the point on the plots where the validation loss is at its lowest and is truly representative of how well the model will perform on independent examples later.

It can be seen from the results that the model with one hidden layer is able to reach a lower validation and **test error rate** in comparison to the two-layer models. From this, we may conclude that this model is the best match for this particular problem. The model with one hidden layer shows a large amount of bias, indicated by the large gap between the training and validation errors, and both were still decreasing, indicating that the model can be trained for more epochs. Lastly, it can be determined from the plots that we should stop training around the region where the validation error rate starts increasing to prevent the model from **overfitting** to the data points.

SUMMARY

In this chapter, you extended your knowledge of deep learning, from understanding the common representations and terminology to implementing them in practice through exercises and activities. You learned how **forward propagation** in neural networks works and how it is used for predicting outputs, how the loss function works as a measure of model performance, and how backpropagation is used to compute the derivatives of loss functions with respect to model parameters.

You also learned about gradient descent, which uses the gradients that are computed by **backpropagation** to gradually update the model parameters. In addition to basic theory and concepts, you implemented and trained both shallow and deep neural networks with Keras and utilized them to make predictions about the output of a given input.

To evaluate your models appropriately, you split a dataset into a training set and a test set as an alternative approach to improving network evaluation and learned the reasons why evaluating a model on training examples can be misleading. This helped further your understanding of overfitting and underfitting that can happen when training a model. Finally, you utilized the training error rate and test error rate to detect overfitting and underfitting in a network and implemented early stopping in order to reduce overfitting in a network.

In the next chapter, you will learn about the Keras wrapper with **scikit-learn** and how to use it to further improve model evaluation by using resampling methods such as cross-validation. By doing this, you will learn how to find the best set of hyperparameters for a deep neural network.

4

EVALUATING YOUR MODEL WITH CROSS-VALIDATION USING KERAS WRAPPERS

OVERVIEW

This chapter introduces you to building Keras wrappers with scikit-learn. You will learn to apply cross-validation to evaluate deep learning models, and create user-defined functions to implement deep learning models along with cross-validation. By the end of this chapter, you will be able to build robust models that perform as well on new, unseen data as they do on the trained data.

INTRODUCTION

In the previous chapter, we experimented with different neural network architectures. We were able to evaluate the performance of the different models by observing the loss and accuracy during the course of the training process. This helped us determine when the model was underfitting or overfitting the training data and how to use techniques such as early stopping to prevent overfitting.

In this chapter, you will learn about **cross-validation**. This is a **resampling technique** that leads to a very accurate and robust estimation of a model's performance, in comparison to the model evaluation approaches we discussed in the previous chapters.

This chapter starts with an in-depth discussion about why we need to use cross-validation for model evaluation, the underlying basics of cross-validation, its variations, and a comparison between them. Next, we will implement cross-validation on Keras deep learning models. We will also use Keras wrappers with scikit-learn to allow Keras models to be treated as estimators in a scikit-learn workflow. You will then learn how to implement cross-validation in scikit-learn and finally bring it all together and perform cross-validation using scikit-learn on Keras deep learning models.

Lastly, you will learn about how to use cross-validation to perform more than just model evaluation, and how a cross-validation estimation of model performance can be used to compare different models and select the one that results in the best performance on a particular dataset. You will also use cross-validation to improve the performance of a given model by finding the best set of hyperparameters for it. We will implement the concepts that we will learn about in this chapter in three activities, each involving a real-life dataset.

CROSS-VALIDATION

Resampling techniques are an important group of techniques in statistical data analysis. They involve repeatedly drawing samples from a dataset to create the training set and the test set. At each repetition, they fit and evaluate the model using the samples drawn from the dataset for the training set and the test set at that repetition.

Using these techniques can provide us with information about the model that is otherwise not obtainable by fitting and evaluating the model only once, using one training set and one test set. Since resampling methods involve fitting a model to the training data several times, they are computationally expensive. Therefore, when it comes to deep learning, we only implement them in the cases where the dataset and the network are relatively small, and the available computational power allows us to do so.

In this section, you will learn about a very important resampling method called **cross-validation**. Cross-validation is one of the most important and most commonly used resampling methods. It computes the best estimation of model performance on new, unseen examples when given a limited dataset. We will also explore the basics of cross-validation, its two variations, and a comparison between them.

DRAWBACKS OF SPLITTING A DATASET ONLY ONCE

In the previous chapter, we mentioned that evaluating a model on the same dataset that's used to train the model is a methodological mistake. Since the model has been trained to reduce the error on this particular set of examples, its performance on it is highly biased. That is why the error rate on training data is always an underestimation of the error rate on new examples. We learned that one way to solve this problem is to randomly hold out a subset of the data as a test set for evaluation and fit the model on the rest of the data, which is called the training set. An illustration of this approach can be seen in the following image:

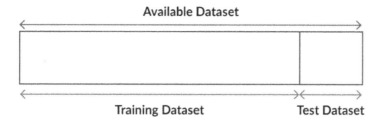

Figure 4.1: Overview of training set/test set split

As we mentioned previously, assigning the data to either the training set or the test set is completely random. This means that if we repeat this process, different data will be assigned to the test set and training set each time. The test error rate that's reported by this approach can vary a lot, depending on which examples are in the test set and which examples are in the training set.

Example

Let's look at an example. Here, we have built a single-layer neural network for the hepatitis C dataset that you saw in *Activity 3.02*, *Advanced Fibrosis Diagnosis with Neural Networks* in *Chapter 3*, *Deep Learning with Keras*. We used the training set/test set approach to compute the test error associated with this model. Instead of splitting and training only once, if we split the data into five separate datasets and repeated this process five times, we might expect five different plots for the test error rates. The test error rates for each of these five experiments can be seen in the following plot:

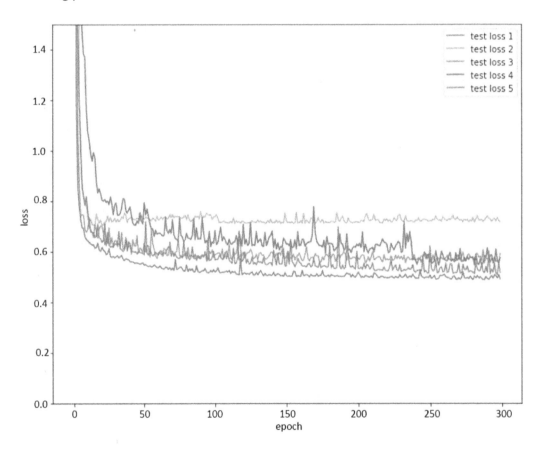

Figure 4.2: Plot of test error rates with five different training set/test set splits on an example dataset

As you can see, the test error rate is quite different in each experiment. This variation in the models' evaluation results indicates that the simple strategy of splitting the dataset into a training set and a test set only once may not lead to a robust and accurate estimation of the model's performance.

To summarize, the training set/test set approach that we learned about in the previous chapter has the obvious advantage of being simple, easy to implement, and computationally inexpensive. However, it has drawbacks too, which are as follows:

- The first drawback is that its estimation of the model's error rate strongly depends on exactly which data is assigned to the test set and which data is assigned to the training set.

- The second drawback is that, in this approach, we are only training the model on a subset of the data. Machine learning models tend to perform worse when they are trained using a small amount of data.

Since the performance of a model can be improved by training it on the entire dataset, we are always looking for ways to include all the available data points in training. Additionally, we are interested in finding a robust estimation of the model's performance by including all the available data points in the evaluation. These objectives can be accomplished with the use of cross-validation techniques. The following are the two methods of cross-validation:

- **K-fold cross-validation**

- **Leave-one-out cross-validation**

K-FOLD CROSS-VALIDATION

In `k-fold cross-validation`, instead of dividing the dataset into two subsets, we divide the dataset into **k** approximately equal-sized subsets or folds. In the first iteration of the method, the first fold is considered a test set. The model is trained on the remaining **k–1** folds, and then it is evaluated on the first fold (the first fold is used to estimate the test error rate).

This process is repeated **k** times, and a different fold is used as the test set in each iteration, while the remaining folds are used as the training set. Eventually, the method results in **k** different test error rates. The final k-fold cross-validation estimate of the model's error rate is computed by averaging these **k** test error rates.

The following diagram illustrates the dataset splitting process in the **k-fold cross-validation** method:

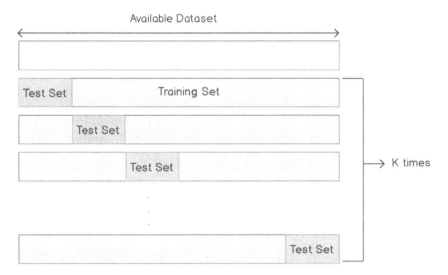

Figure 4.3: Overview of dataset splitting in the k-fold cross-validation method

In practice, we usually perform **k-fold cross-validation** with **k=5** or **k=10**, and these are the recommended values if you are struggling to select a value for your dataset. Deciding on the number of folds to use is dependent on the number of examples in the dataset and the available computational power. If **k=5**, the model will be trained and evaluated five times, while if **k=10**, this process will be repeated 10 times. The higher the number of folds, the longer it will take to perform k-fold cross-validation.

In k-fold cross-validation, the assignment of examples to each fold is completely random. However, by looking at the preceding diagram, you will see that, in the end, every single piece of data is used for both training and evaluation. That's why if you repeat k-fold cross-validation many times on the same dataset and the same model, the final reported test error rates will be almost identical. Therefore, k-fold cross-validation does not suffer from high variance in its results, in contrast to the training set/test set approach. Now, we will take a look at the second form of cross-validation: leave-one-out validation.

LEAVE-ONE-OUT CROSS-VALIDATION

Leave-One-Out (LOO) is a variation of the cross-validation technique in which, instead of dividing the dataset into two comparable-sized subsets for the training set and test set, only one single piece of data is used for evaluation. If there are **n** data examples in the entire dataset, at each iteration of **LOO cross-validation**, the model is trained on **n-1** examples and the single remaining example is used to compute the test error rate.

Using only one example for estimating the test error rate leads to an unbiased but high variance estimation of model performance; it is unbiased because this one example has not been used in training the model, it has high variance because it is computed based on only one data example, and it will vary depending on which exact data example is used. This process is repeated **n** times, and, at each iteration, a different data example is used for evaluation. In the end, the method will result in **n** different test error rates, and the final **LOO cross-validation** test error estimation is computed by averaging these **n** error rates.

An illustration of the dataset splitting process in the **LOO cross-validation** method can be seen in the following diagram:

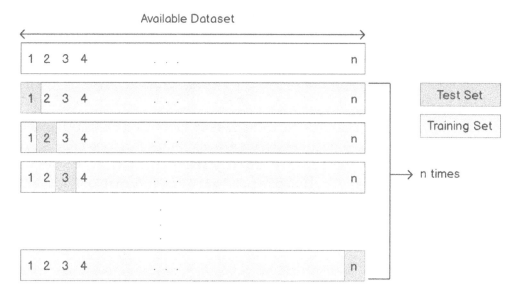

Figure 4.4: Overview of dataset splitting in the LOO cross-validation method

In each iteration of **LOO cross-validation**, almost all the examples in the dataset are used to train the model. On the other hand, in the training set/test set approach, a relatively large subset of data is used for evaluation and not used in training. Therefore, the **LOO** estimation of model performance is much closer to the performance of a model that is trained on the entire dataset, and this is the main advantage of **LOO cross-validation** over the **training set/test set** approach.

Additionally, since in each iteration of **LOO cross-validation** only one unique data example is used for evaluation, and every single data example is used for training as well, there is no randomness associated with this method. Therefore, if you repeat **LOO cross-validation** many times on the same dataset and the same model, the final reported test error rates will be exactly the same each time.

The drawback of **LOO cross-validation** is that it is computationally expensive. The reason for this is that the model needs to be trained **n** times, and in cases where **n** is large and/or the network is large, it will take a long time to complete. Both **LOO** and **k-fold cross-validation** have their advantages and disadvantages, all of which we will compare in the next section.

COMPARING THE K-FOLD AND LOO METHODS

By comparing the two preceding diagrams, it is obvious that **LOO cross-validation** is, in fact, a special case of **k-fold cross-validation**, where **k=n**. However, as was mentioned previously, choosing **k=n** is computationally very expensive in comparison to choosing **k=5** or **k=10**.

Therefore, the first advantage of **k-fold cross-validation** over **LOO cross-validation** is that it is computationally less expensive. The following table compares the **k-fold with low-k**, **k-fold with high-k** and **LOO**, and **no cross-validation** with respect to **bias** and **variance**. The table shows that the highest bias comes with a simple **train-test split approach** and that the highest variance comes with leave-one-put cross-validation. In the middle is **k-fold cross-validation**. This is why k-fold cross-validation is generally the most appropriate choice for most machine learning tasks:

	Bias	Variance
Simple train-test split approach	Highest	Lowest
K-fold cross-validation with smaller k	Higher	Lower
K-fold cross-validation with larger k	Lower	Higher
LOO cross-validation k=n	Lowest	Highest

Figure 4.5: Comparing the train-test split, k-fold cross-validation, and LOO cross-validation methods

The following plot compares the **training set/test set** approach, **k-fold cross-validation**, and **LOO cross-validation** in terms of **bias** and **variance**:

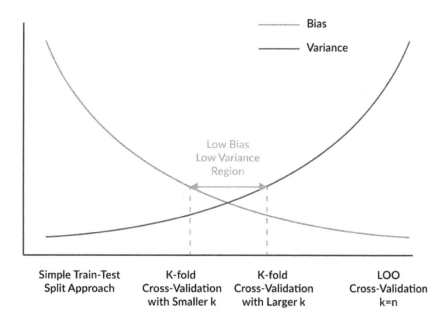

Figure 4.6: Comparing the training set/test set approach, k-fold cross-validation, and LOO cross-validation in terms of bias and variance

Generally, in machine learning and data analysis, the most desirable model is the one with the **lowest bias** and the **lowest variance**. As shown in the preceding plot, the region labeled in the middle of the graph, where both **bias** and **variance** are low, is of interest. It turns out that this region is equivalent to **k-fold cross-validation** with **k** between **5** and **10**. In the next section, we will explore how to implement various methods of cross-validation in practice.

CROSS-VALIDATION FOR DEEP LEARNING MODELS

In this section, you will learn about using the Keras wrapper with scikit-learn, which is a helpful tool that allows us to use Keras models as part of a scikit-learn workflow. As a result, scikit-learn methods and functions, such as the one for performing cross-validation, can easily be applied to Keras models.

You will learn, step-by-step, how to implement what you learned about cross-validation in the previous section using scikit-learn. Furthermore, you will learn how to use cross-validation to evaluate Keras deep learning models using the Keras wrapper with scikit-learn. Lastly, you will practice what you have learned by solving a problem involving a real dataset.

KERAS WRAPPER WITH SCIKIT-LEARN

When it comes to general machine learning and data analysis, the scikit-learn library is much richer and easier to use than Keras. That is why being able to use scikit-learn methods on Keras models will be of great value.

Fortunately, Keras comes with a helpful wrapper, **keras.wrappers.scikit_learn**, that allows us to build scikit-learn interfaces for deep learning models that can be used as classification or regression estimators in scikit-learn. There are two types of wrapper: one for classification estimators and one for regression estimators. The following code is used to define these scikit-learn interfaces:

```
keras.wrappers.scikit_learn.KerasClassifier(build_fn=None, **sk_params)
# wrappers for classification estimators
keras.wrappers.scikit_learn.KerasRegressor(build_fn=None, **sk_params)
# wrappers for regression estimators
```

The **build_fn** argument needs to be a callable function where a Keras sequential model is defined, compiled and returned inside its body.

The **sk_params** argument can take parameters for building the model (such as activation functions for layers) and parameters for fitting the model (such as the number of epochs and batch size). This will be put into practice in the following exercise, where we will use Keras wrappers for a regression problem.

> **NOTE**
>
> All the activities in this chapter will be developed in a Jupyter notebook. Please download this book's GitHub repository, along with all the prepared templates, which can be found here:
>
> https://packt.live/3btnjfA.

EXERCISE 4.01: BUILDING THE KERAS WRAPPER WITH SCIKIT-LEARN FOR A REGRESSION PROBLEM

In this exercise, you will learn the step-by-step process of building the wrapper for a Keras deep learning model so that it can be used in a scikit-learn workflow. First, load in the dataset of **908** data points of a regression problem, where each record describes six attributes of a chemical, and the target is the acute toxicity toward the fish Pimephales promelas, or **LC50**:

> **NOTE**
>
> Watch out for the slashes in the string below. Remember that the backslashes (****) are used to split the code across multiple lines, while the forward slashes (**/**) are part of the path.

```
# import data
import pandas as pd

colnames = ['CICO', 'SM1_Dz(Z)', 'GATS1i', \
            'NdsCH', 'NdssC','MLOGP', 'LC50']
data = pd.read_csv('../data/qsar_fish_toxicity.csv', \
                   sep=';', names=colnames)
X = data.drop('LC50', axis=1)
y = data['LC50']

# Print the sizes of the dataset
```

```
print("Number of Examples in the Dataset = ", X.shape[0])
print("Number of Features for each example = ", X.shape[1])
# print output range
print("Output Range = [%f, %f]" % (min(y), max(y)))
```

This is the expected output:

```
Number of Examples in the Dataset =  908
Number of Features for each example =  6
Output Range = [0.053000, 9.612000]
```

Since the output in this dataset takes a numerical value, this is a regression problem. The goal is to build a model that predicts the acute toxicity toward the fish **LC50**, given the other attributes of the chemical. Now, let's go through the steps:

1. Define a function that builds and returns a Keras model for this regression problem. The Keras model that you define must have a single hidden layer of size **8** with **ReLU activation** functions. Also, use the **Mean Squared Error (MSE)** loss function and the **Adam optimizer** to compile the model:

```
from keras.models import Sequential
from keras.layers import Dense, Activation
# Create the function that returns the keras model
def build_model():
    # build the Keras model
    model = Sequential()
    model.add(Dense(8, input_dim=X.shape[1], \
            activation='relu'))
    model.add(Dense(1))
    # Compile the model
    model.compile(loss='mean_squared_error', \
                optimizer='adam')
    # return the model
    return model
```

2. Now, use the Keras wrapper with scikit-learn to create the scikit-learn interface for your model. Remember that you need to provide the **epochs**, **batch_size**, and **verbose** arguments here:

```
# build the scikit-Learn interface for the keras model
from keras.wrappers.scikit_learn import KerasRegressor
YourModel = KerasRegressor(build_fn= build_model, \
                           epochs=100, \
                           batch_size=20, \
                           verbose=1)
```

Now, **YourModel** is ready to be used as a regression estimator in scikit-learn.

In this exercise, we learned how to build a Keras wrapper with scikit-learn for a regression problem by using a simulated dataset.

> **NOTE**
>
> To access the source code for this specific section, please refer to https://packt.live/38nuqVP.
>
> You can also run this example online at https://packt.live/31MLgMF.

We will continue implementing cross-validation using this dataset in the rest of the exercises in this chapter.

CROSS-VALIDATION WITH SCIKIT-LEARN

In the previous chapter, you learned that you can perform **training set/test set** splitting easily in scikit-learn. Let's assume that your original dataset is stored in **X** and **y** arrays. You can split them randomly into a training set and a test set using the following commands:

```
from sklearn.model_selection import train_test_split
X_train, X_test, y_train, y_test = train_test_split\
                              (X, y, test_size=0.3, \
                               random_state=0)
```

The **test_size** argument can be assigned to any number between **0** and **1**, depending on how large you would like the test set to be. By providing an **int** number for a **random_state** argument, you will be able to select the seed for the random number generator.

The easiest way to perform cross-validation in scikit-learn is by using the **cross_val_score** function. In order to do this, you need to define your estimator first (in our case, the estimator will be a Keras model). Then, you will be able to perform cross-validation on your estimator/model using the following commands:

```
from sklearn.model_selection import cross_val_score
scores = cross_val_score(YourModel, X, y, cv=5)
```

Notice that we provide the Keras model and the original dataset as arguments to the **cross_val_score** function, along with the number of folds (the **cv** argument). Here, we used **cv=5**, so the **cross_val_score** function will randomly split the dataset into five-folds and perform training and fitting on the model five times using five different training and test sets. It will compute the default metric for model evaluation (or the metrics given to the Keras model when defining it) at each iteration/fold and store them in scores. We can print the final cross-validation score as follows:

```
print(scores.mean())
```

Earlier, we mentioned that the score that's returned by the **cross_val_score** function is the default metric for our model or the metric that we determined for it when defining our model. However, it is possible to change the cross-validation metric by providing the desired metric as a **scoring** argument when calling the **cross_val_score** function.

> **NOTE**
>
> You can learn more about how to provide the desired metric in the **scoring** argument of the **cross_val_score** function here: https:// scikit-learn.org/stable/modules/model_evaluation.html#scoring-parameter.

By providing an integer number for the **cv** argument of the **cross_val_score** function, we are telling the function to perform k-fold cross-validation on the dataset. However, there are several other iterators available in scikit-learn that we can assign to **cv** to perform other variations of cross-validation on the dataset. For example, the following code block will perform **LOO cross-validation** on the dataset:

```
from sklearn.model_selection import LeaveOneOut
loo = LeaveOneOut()
scores = cross_val_score(YourModel, X, y, cv=loo)
```

In the next section, we will explore k-fold cross-validation in scikit-learn and see how it can be used with Keras models.

CROSS-VALIDATION ITERATORS IN SCIKIT-LEARN

A list of the most commonly used cross-validation iterators available in scikit-learn is provided here, along with a brief description of each of them:

- **KFold(n_splits=?)**

  This divides the dataset into k folds or groups. The **n_splits** argument is required to determine how many folds to use. If **n_splits=n**, it will be equivalent to **LOO cross-validation**.

- **RepeatedKFold(n_splits=?, n_repeats=?, random_ state=random_state)**

  This will repeat k-fold cross-validation **n_repeats** times.

- **LeaveOneOut()**

  This will split the dataset for **LOO cross-validation**.

- **ShuffleSplit(n_splits=?, test_size=?, random_state= random_state)**

  This will generate an **n_splits** number of random and independent training set/test set dataset splits. It is possible to store the seed for the random number generator using the **random_state** argument; if you do this, the dataset splits will be reproducible.

In addition to the regular iterators, such as the ones mentioned here, there are **stratified** versions as well. Stratified sampling is useful when the number of examples in different classes of a dataset is unbalanced. For example, imagine that we want to design a classifier to predict whether someone will default on their credit card debt, where almost **95%** of the examples in the dataset are in the **negative** class. Stratified sampling makes sure that the relative class frequencies are preserved in each **training set/test set** split. It is recommended to use the stratified versions of iterators for such cases.

Usually, before using a training set to train and evaluate a model, we perform preprocessing on it to scale the examples so that they have a mean equal to **0** and a standard deviation equal to **1**. In the **training set/test set** approach, we need to scale the training set and store the transformation. The following code block will do this for us:

```
from sklearn.preprocessing import StandardScaler
scaler = StandardScaler()
X_train = scaler.fit_transform(X_train)
X_test = scaler.transform(X_test)
```

Here's an example of performing **stratified k-fold cross-validation** with **k=5** on our **X**, **y** dataset:

```
from sklearn.model_selection import StratifiedKFold
skf = StratifiedKFold(n_splits=5)
scores = cross_val_score(YourModel, X, y, cv=skf)
```

> **NOTE**
>
> You can learn more about cross-validation iterators in scikit-learn here:
>
> https://scikit-learn.org/stable/modules/cross_validation.html#cross-validation-iterators.

Now that we understand cross-validation iterators, we can put them into practice in an exercise.

EXERCISE 4.02: EVALUATING DEEP NEURAL NETWORKS WITH CROSS-VALIDATION

In this exercise, we will bring all the concepts and methods that we have learned about in this topic about cross-validation together. We will go through all the steps one more time, from defining a Keras deep learning model to transferring it to a scikit-learn workflow and performing cross-validation in order to evaluate its performance. In a sense, this exercise is a recap of what we have learned so far, and what is covered here will be extremely helpful for *Activity 4.01, Model Evaluation Using Cross-Validation for an Advanced Fibrosis Diagnosis Classifier*:

1. The first step is always to load the dataset that you would like to build the model for. First, load in the dataset of **908** data points of a regression problem, where each record describes six attributes of a chemical and the target is the acute toxicity toward the fish Pimephales promelas, or **LC50**:

```
# import data
import pandas as pd

colnames = ['CIC0', 'SM1_Dz(Z)', 'GATS1i', \
            'NdsCH', 'NdssC','MLOGP', 'LC50']
data = pd.read_csv('../data/qsar_fish_toxicity.csv', \
                   sep=';', names=colnames)
X = data.drop('LC50', axis=1)
y = data['LC50']

# Print the sizes of the dataset
print("Number of Examples in the Dataset = ", X.shape[0])
print("Number of Features for each example = ", X.shape[1])
# print output range
print("Output Range = [%f, %f]" % (min(y), max(y)))
```

The output is as follows:

```
Number of Examples in the Dataset =  908
Number of Features for each example =  6
Output Range = [0.053000, 9.612000]
```

2. Define the function that returns the Keras model with a single hidden layer of size **8** with **ReLU activation** functions using the **Mean Squared Error (MSE)** loss function and the **Adam optimizer**:

```
from keras.models import Sequential
from keras.layers import Dense, Activation
# Create the function that returns the keras model
def build_model():
    # build the Keras model
    model = Sequential()
    model.add(Dense(8, input_dim=X.shape[1], \
            activation='relu'))
    model.add(Dense(1))
    # Compile the model
    model.compile(loss='mean_squared_error', \
                optimizer='adam')
    # return the model
    return model
```

3. Set the **seed** and use the wrapper to build the scikit-learn interface for the Keras model we defined in the function in **step 2**:

```
# build the scikit-Learn interface for the keras model
from keras.wrappers.scikit_learn import KerasRegressor
import numpy as np
from tensorflow import random
seed = 1
np.random.seed(seed)
random.set_seed(seed)
YourModel = KerasRegressor(build_fn= build_model, \
                        epochs=100, batch_size=20, \
                        verbose=1 , shuffle=False)
```

4. Define the iterator to use for cross-validation. Let's perform **5-fold cross-validation**:

```
# define the iterator to perform 5-fold cross-validation
from sklearn.model_selection import KFold
kf = KFold(n_splits=5)
```

5. Call the **cross_val_score** function to perform cross-validation. This step will take a while to complete, depending on the computational power that's available:

```
# perform cross-validation on X, y
from sklearn.model_selection import cross_val_score
results = cross_val_score(YourModel, X, y, cv=kf)
```

6. Once cross-validation has been completed, print the **final cross-validation** estimation of model performance (the default metric for performance will be the test loss):

```
# print the result
print(f"Final Cross-Validation Loss = {abs(results.mean()):.4f}")
```

Here's an example output:

```
Final Cross-Validation Loss = 0.9680
```

The cross-validation loss states that the Keras model that was trained on this dataset is able to predict the **LC50** of the chemicals with an average loss of **0.9680**. We will try to examine this model further in the next exercise.

These were all the steps that are required in order to evaluate a Keras deep learning model using cross-validation in scikit-learn. Now, we will put them into practice in an activity.

> **NOTE**
>
> To access the source code for this specific section, please refer to https://packt.live/3eRTITM.
>
> You can also run this example online at https://packt.live/31IdVT0.

ACTIVITY 4.01: MODEL EVALUATION USING CROSS-VALIDATION FOR AN ADVANCED FIBROSIS DIAGNOSIS CLASSIFIER

We learned about the hepatitis C dataset in *Activity 3.02, Advanced Fibrosis Diagnosis with Neural Networks* of *Chapter 3, Deep Learning with Keras*. The dataset consists of information for **1385** patients who underwent treatment dosages for hepatitis C. For each patient, **28** different attributes are available, such as age, gender, and BMI, as well as a class label, which can only take two values: **1**, indicating advanced fibrosis, and **0**, indicating no indication of advanced fibrosis. This is a **binary/two-class** classification problem with an input dimension equal to **28**.

In *Chapter 3, Deep Learning with Keras*, we built Keras models to perform classification on this dataset. We trained and evaluated the models using **training set/test set** splitting and reported the test error rate. In this activity, we are going to use what we learned in this topic to train and evaluate a deep learning model using **k-fold cross-validation**. We will use the model that resulted in the best test error rate from the previous activity. The goal is to compare the cross-validation error rate with the training set/test set approach error rate:

1. Import the necessary libraries. Load the dataset from the **data** subfolder of the **Chapter04** folder from GitHub using **X = pd.read_csv('../data/HCV_feats.csv'), y = pd.read_csv('../data/HCV_target.csv')**. Print the number of examples in the dataset, the number of features available, and the possible values for the class labels.

2. Define the function that returns the Keras model. The Keras model will be a deep neural network with two hidden layers, where the **first hidden layer** is of **size 4** and the **second hidden layer** is of **size 2**, and use the **tanh activation** function to perform the classification. Use the following values for the hyperparameters:

   ```
   optimizer = 'adam', loss = 'binary_crossentropy', metrics
   = ['accuracy']
   ```

3. Build the scikit-learn interface for the Keras model with **epochs=100, batch_size=20**, and **shuffle=False**. Define the cross-validation iterator as **StratifiedKFold** with **k=5**. Perform k-fold cross-validation on the model and store the scores.

4. Print the accuracy for each iteration/fold, plus the overall cross-validation accuracy and its associated standard deviation.

5. Compare this result with the result from *Activity 3.02, Advanced Fibrosis Diagnosis with Neural Networks* of *Chapter 3, Deep Learning with Keras*.

After implementing the preceding steps, the expected output will look as follows:

```
Test accuracy at fold 1 = 0.5198556184768677
Test accuracy at fold 2 = 0.4693140685558319
Test accuracy at fold 3 = 0.512635350227356
Test accuracy at fold 4 = 0.5740072131156921
Test accuracy at fold 5 = 0.5523465871810913

Final Cross Validation Test Accuracy: 0.5256317675113678
Standard Deviation of Final Test Accuracy: 0.03584760640500936
```

> **NOTE**
>
> The solution for this activity can be found on page 381.

The accuracy we received from training set/test set approach we performed in *Activity 3.02, Advanced Fibrosis Diagnosis with Neural Networks* of *Chapter 3, Deep Learning with Keras*, was **49.819%**, which is lower than the test accuracy we achieved when performing **5-fold cross-validation** on the same deep learning model and the same dataset, but lower than the accuracy on one of the folds.

The reason for this difference is that the test error rate resulting from the training set/test set approach was computed by only including a subset of the data points in the model's evaluation. On the other hand, the test error rate here is computed by including all the data points in the evaluation, and therefore this estimation of the model's performance is more accurate and more robust, performing better on the unseen test dataset.

In this activity, we used cross-validation to perform a model evaluation on a problem involving a real dataset. Improving model evaluation is not the only purpose of using cross-validation, and it can be used to select the best model or parameters for a given problem as well.

MODEL SELECTION WITH CROSS-VALIDATION

Cross-validation provides us with robust estimation of model performance on unseen examples. For this reason, it can be used to decide between two models for a particular problem or to decide which model parameters (or hyperparameters) to use for a particular problem. In these cases, we would like to find out which model or which set of model parameters/hyperparameters results in the lowest test error rate. Therefore, we will select that model or that set of parameters/hyperparameters for our problem.

In this section, you are going to practice using cross-validation for this purpose. You will learn how to define a set of hyperparameters for your deep learning model and then write user-defined functions in order to perform cross-validation on your model for each of the possible combinations of hyperparameters. Then, you will observe which combination of hyperparameters leads to the lowest test error rate, and that combination will be your choice for your final model.

CROSS-VALIDATION FOR MODEL EVALUATION VERSUS MODEL SELECTION

In this section, we are going to go deeper into what it means to use cross-validation for model evaluation versus model selection. So far, we have learned that evaluating a model on the training set results in an underestimation of the model's error rate on unseen examples. Splitting the dataset into a **training set** and a **test set** gives us a more accurate estimation of the model's performance but suffers from high variance.

Lastly, cross-validation results in a much more robust and accurate estimation of the model's performance on unseen examples. An illustration of the error rate estimations resulting from these three approaches for model evaluation can be seen in the following plot.

The following plot shows the case where the error rate estimation in the training set/ test set approach is slightly lower than the cross-validation estimation. However, it is important to remember that the **training set/test set** error rate can be higher than the cross-validation estimation error rate as well, depending on what data is included in the test set (hence the high variance problem). On the other hand, the error rate resulting from performing an evaluation on the training set is always lower than the other two approaches:

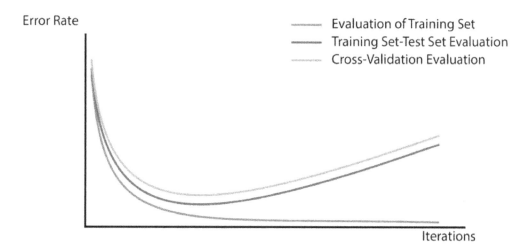

Figure 4.7: Illustration of the error rate estimations resulting from the three approaches to model evaluation

We have established that cross-validation leads to the best estimation of a model's performance on independent data examples. Knowing this, we can use cross-validation to decide which model to use for a particular problem. For example, if we have four different models and we would like to decide which one is a better fit for a particular dataset, we can train and evaluate each of the four models using cross-validation and choose the model with the lowest cross-validation error rate as our final model for the dataset. The following plot shows the cross-validation error rate associated with four hypothetical models. From this, we can conclude that **Model 1** is the best fit for the problem, while **Model 4** is the worst choice. These four models could be deep neural networks with a different number of hidden layers and a different number of units in their hidden layers:

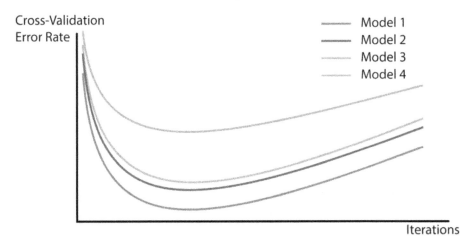

Figure 4.8: Illustration of cross-validation error rates associated with four hypothetical models

After we have found out which model is the best fit for a particular problem, the next step is choosing the best set of parameters or hyperparameters for that model. Previously, we discussed that when building a deep neural network, several hyperparameters need to be selected for the model, and several choices are available for each of these hyperparameters.

These hyperparameters include the type of activation function, loss function, and optimizer, plus the number of epochs and batch size. We can define the set of possible choices for each of these hyperparameters and then implement the model, along with cross-validation, to find the best combination of hyperparameters.

An illustration of the cross-validation error rates associated with four different sets of hyperparameters for a hypothetical deep learning model is shown in the following plot. From this, we can conclude that **Set 1** is the best choice for this model since the line corresponding to **Hyperparameters Set 1** has the lowest value for the cross-validation error rate:

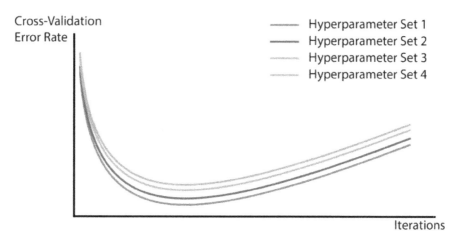

Figure 4.9: Illustration of cross-validation error rates associated with four different sets of hyperparameters for a hypothetical deep learning model

In the next exercise, we will learn how to iterate through different model architectures and hyperparameters to find the set that results in an optimal model.

EXERCISE 4.03: WRITING USER-DEFINED FUNCTIONS TO IMPLEMENT DEEP LEARNING MODELS WITH CROSS-VALIDATION

In this exercise, you will learn how to use cross-validation for the purpose of model selection.

First, load in the dataset of **908** data points of a regression problem, where each record describes six attributes of a chemical and the target is the acute toxicity toward the fish Pimephales promelas, or **LC50**. The goal is to build a model to predict the **LC50** of each chemical, given the chemical attributes:

```
# import data
import pandas as pd
import numpy as np
from tensorflow import random
```

```
colnames = ['CIC0', 'SM1_Dz(Z)', 'GATS1i', 'NdsCH', \
            'NdssC','MLOGP', 'LC50']
data = pd.read_csv('../data/qsar_fish_toxicity.csv', \
                    sep=';', names=colnames)
X = data.drop('LC50', axis=1)
y = data['LC50']
```

Follow these steps to complete this exercise:

1. Define three functions to return three Keras models. The first model should have one hidden layer of **size 4**, the second model should have one hidden layer of **size 8**, and the third model should have two hidden layers, with the first layer of **size 4** and the second layer of **size 2**. Use a **ReLU activation** function for all the hidden layers. The goal is to find out which of these three models leads to the lowest cross-validation error rate:

```
# Define the Keras models
from keras.models import Sequential
from keras.layers import Dense

def build_model_1():
    # build the Keras model_1
    model = Sequential()
    model.add(Dense(4, input_dim=X.shape[1], \
                    activation='relu'))
    model.add(Dense(1))
    # Compile the model
    model.compile(loss='mean_squared_error', \
                  optimizer='adam')
    # return the model
    return model

def build_model_2():
    # build the Keras model_2
    model = Sequential()
    model.add(Dense(8, input_dim=X.shape[1], \
            activation='relu'))
    model.add(Dense(1))
    # Compile the model
    model.compile(loss='mean_squared_error', \
                  optimizer='adam')
    # return the model
```

```
        return model

def build_model_3():
    # build the Keras model_3
    model = Sequential()
    model.add(Dense(4, input_dim=X.shape[1], \
                    activation='relu'))
    model.add(Dense(2, activation='relu'))
    model.add(Dense(1))
    # Compile the model
    model.compile(loss='mean_squared_error', \
                  optimizer='adam')
    # return the model
    return model
```

2. Write a loop to build the Keras wrapper and perform **3-fold cross-validation** on the **three** models. Store the scores for each model:

```
"""
define a seed for random number generator so the result will be
reproducible
"""
seed = 1
np.random.seed(seed)
random.set_seed(seed)
# perform cross-validation on each model
from keras.wrappers.scikit_learn import KerasRegressor
from sklearn.model_selection import KFold
from sklearn.model_selection import cross_val_score
results_1 = []
models = [build_model_1, build_model_2, build_model_3]
# loop over three models
for m in range(len(models)):
    model = KerasRegressor(build_fn=models[m], \
                           epochs=100, batch_size=20, \
                           verbose=0, shuffle=False)
    kf = KFold(n_splits=3)
    result = cross_val_score(model, X, y, cv=kf)
    results_1.append(result)
```

3. Print the final cross-validation error rate for each of the models to find out which model has a lower error rate:

```
# print the cross-validation scores
print("Cross-Validation Loss for Model 1 =", \
     abs(results_1[0].mean()))
print("Cross-Validation Loss for Model 2 =", \
     abs(results_1[1].mean()))
print("Cross-Validation Loss for Model 3 =", \
     abs(results_1[2].mean()))
```

Here's an example output:

```
Cross-Validation Loss for Model 1 = 0.990475798256843
Cross-Validation Loss for Model 2 = 0.926532513151634
Cross-Validation Loss for Model 3 = 0.9735719371528117
```

Model 2 results in the lowest error rate, so we will use it in the steps that follow.

4. Use cross-validation again to determine the number of epochs and batch size for the model that resulted in the lowest cross-validation error rate. Write the code that performs **3-fold cross-validation** on every possible combination of **epochs** and **batch-size** in the ranges **epochs=[100, 150]** and **batch_size=[20, 15]** and store the scores:

```
"""
define a seed for random number generator so the result will be
reproducible
"""
np.random.seed(seed)
random.set_seed(seed)
results_2 = []
epochs = [100, 150]
batches = [20, 15]
# Loop over pairs of epochs and batch_size
for e in range(len(epochs)):
    for b in range(len(batches)):
        model = KerasRegressor(build_fn= build_model_3, \
                               epochs= epochs[e], \
                               batch_size= batches[b], \
                               verbose=0, \
                               shuffle=False)
```

```
kf = KFold(n_splits=3)
result = cross_val_score(model, X, y, cv=kf)
results_2.append(result)
```

> **NOTE**
>
> The preceding code block uses two **for** loops to perform **3-fold cross-validation** for all possible combinations of **epochs** and **batch_size**. Since there are two choices for each of them, four different pairs are possible and therefore cross-validation will be performed four times.

5. Print the final cross-validation error rate for each of the **epochs/batch_size** pairs to find out which pair has the lowest error rate:

```
"""
Print cross-validation score for each possible pair of epochs, batch_
size
"""
c = 0
for e in range(len(epochs)):
    for b in range(len(batches)):
        print("batch_size =", batches[b],", \
            epochs =", epochs[e], ", Test Loss =", \
            abs(results_2[c].mean()))
        c += 1
```

Here's an example output:

```
batch_size = 20 , epochs = 100 , Test Loss = 0.9359159401008821
batch_size = 15 , epochs = 100 , Test Loss = 0.9642481369794683
batch_size = 20 , epochs = 150 , Test Loss = 0.9561188386646661
batch_size = 15 , epochs = 150 , Test Loss = 0.9359079093029896
```

As you can see, the performance for **epochs=150** and **batch_size=15**, and for **epochs=100** and **batch_size=20**, are almost the same. Therefore, we will choose **epochs=100** and **batch_size=20** in the next step to speed up this process.

6. Use cross-validation again in order to decide on the activation function for the hidden layers and the optimizer for the model from **activations = ['relu', 'tanh']** and **optimizers = ['sgd', 'adam', 'rmsprop']**. Remember to use the best pair of **batch_size** and **epochs** from the previous step:

```
# Modify build_model_2 function
def build_model_2(activation='relu', optimizer='adam'):
    # build the Keras model_2
    model = Sequential()
    model.add(Dense(8, input_dim=X.shape[1], \
            activation=activation))
    model.add(Dense(1))
    # Compile the model
    model.compile(loss='mean_squared_error', \
                optimizer=optimizer)
    # return the model
    return model


results_3 = []
activations = ['relu', 'tanh']
optimizers = ['sgd', 'adam', 'rmsprop']


"""
Define a seed for the random number generator so the result will be
reproducible
"""
np.random.seed(seed)
random.set_seed(seed)
# Loop over pairs of activation and optimizer
for o in range(len(optimizers)):
    for a in range(len(activations)):
        optimizer = optimizers[o]
        activation = activations[a]
        model = KerasRegressor(build_fn= build_model_3, \
                            epochs=100, batch_size=20, \
                            verbose=0, shuffle=False)
        kf = KFold(n_splits=3)
        result = cross_val_score(model, X, y, cv=kf)
        results_3.append(result)
```

> **NOTE**
>
> Notice that we had to modify the **build_model_2** function by passing the **activation**, the **optimizer**, and their default values as arguments of the function.

7. Print the final cross-validation error rate for each pair of **activation** and **optimizer** to find out which pair has the lower error rate:

```
"""
Print cross-validation score for each possible pair of optimizer,
activation
"""
c = 0
for o in range(len(optimizers)):
    for a in range(len(activations)):
        print("activation = ", activations[a],", \
                optimizer = ", optimizers[o], ", \
                Test Loss = ", abs(results_3[c].mean()))
        c += 1
```

Here's the output:

```
activation =  relu , optimizer =  sgd , Test Loss =
1.0123592540516995
activation =  tanh , optimizer =  sgd , Test Loss =
3.393908379781118
activation =  relu , optimizer =  adam , Test Loss =
0.9662686089392641
activation =  tanh , optimizer =  adam , Test Loss =
2.1369285960222144
activation =  relu , optimizer =  rmsprop , Test Loss =
2.1892826984214984
activation =  tanh , optimizer =  rmsprop , Test Loss =
2.2029884275363014
```

8. The **activation='relu'** and **optimizer='adam'** pair results in the lowest error rate. Also, the result for the **activation='relu'** and **optimizer='sgd'** pair is almost as good. Therefore, we can use either of these optimizers in the final model to predict the aquatic toxicity for this dataset.

> **NOTE**
>
> To access the source code for this specific section, please refer to https://packt.live/2BYCwbg.
>
> You can also run this example online at https://packt.live/3gofLfP.

Now, you are ready to practice model selection using cross-validation on another dataset. In *Activity 4.02, Model Selection Using Cross-Validation for the Advanced Fibrosis Diagnosis Classifier*, you will practice these steps further by implementing them by yourself on a classification problem with the hepatitis C dataset.

> **NOTE**
>
> *Exercise 4.02, Evaluating Deep Neural Networks with Cross-Validation*, and *Exercise 4.03, Writing User-Defined Functions to Implement Deep Learning Models with Cross-Validation*, involve performing **k-fold cross-validation** several times, so the steps may take several minutes to complete. If they are taking too long to complete, you may want to try speeding up the process by decreasing the number of folds or epochs or increasing the batch sizes. Obviously, if you do so, you will get different results compared to the expected outputs, but the same principles still apply for selecting the model and hyperparameters.

ACTIVITY 4.02: MODEL SELECTION USING CROSS-VALIDATION FOR THE ADVANCED FIBROSIS DIAGNOSIS CLASSIFIER

In this activity, we are going to improve our classifier for the hepatitis C dataset by using cross-validation for model selection and hyperparameter selection. Follow these steps:

1. Import the required packages. Load the dataset from the **data** subfolder of the **Chapter04** folder from GitHub using **X = pd.read_csv('../data/HCV_feats.csv'), y = pd.read_csv('../data/HCV_target.csv')**.

2. Define three functions, each returning a different Keras model. The first Keras model will be a deep neural network with three hidden layers all of **size 4** and **ReLU activation** functions. The second Keras model will be a deep neural network with two hidden layers, the first layer of **size 4** and the second later of **size 2**, and **ReLU activation** functions. The third Keras model will be a deep neural network with two hidden layers, both of **size 8**, and a **ReLU activation** function. Use the following values for the hyperparameters:

   optimizer = 'adam', loss = 'binary_crossentropy', metrics = ['accuracy']

3. Write the code that will loop over the three models and perform **5-fold cross-validation** on each of them (use **epochs=100, batch_size=20**, and **shuffle=False** in this step). Store all the cross-validation scores in a list and print the results. Which model results in the best accuracy?

   > **NOTE**
   >
   > *Steps 3, 4,* and *5* of this activity involve performing **5-fold cross-validation** three, four, and six times, respectively. Therefore, they may take some time to complete.

4. Write the code that uses the **epochs = [100, 200]** and **batches = [10, 20]** values for **epochs** and **batch_size**. Perform 5-fold cross-validation for each possible pair on the Keras model that resulted in the best accuracy from *step 3*. Store all the cross-validation scores in a list and print the results. Which **epochs** and **batch_size** pair results in the best accuracy?

5. Write the code that uses the **optimizers = ['rmsprop', 'adam','sgd']** and **activations = ['relu', 'tanh']** values for **optimizer** and **activation**. Perform 5-fold cross-validation for each possible pair on the Keras model that resulted in the best accuracy from *step 3*. Use the **batch_size** and **epochs** values that resulted in the best accuracy from *step 4*. Store all the cross-validation scores in a list and print the results. Which **optimizer** and **activation** pair results in the best accuracy?

> **NOTE**
>
> Please note that there is randomness associated with initializing weights and biases in a deep neural network, as well as with selecting which examples to include in each fold when performing k-fold cross-validation. Therefore, you might get a completely different result if you run the exact same code twice. For this reason, it is important to set up seeds when building and training neural networks, as well as when performing cross-validation. By doing this, you can make sure that you are repeating the exact same neural network initialization and the exact same training sets and test sets when you rerun the code.

After implementing these steps, the expected output will be as follows:

```
activation =   relu , optimizer =   rmsprop , Test accuracy =
0.5234657049179077
activation =   tanh , optimizer =   rmsprop , Test accuracy =
0.49602887630462644
activation =   relu , optimizer =   adam , Test accuracy =
0.5039711117744445
activation =   tanh , optimizer =   adam , Test accuracy =
0.4989169597625732
activation =   relu , optimizer =   sgd , Test accuracy =
0.48953068256378174
activation =   tanh , optimizer =   sgd , Test accuracy =
0.5191335678100586
```

> **NOTE**
>
> The solution for this activity can be found on page 384.

In this activity, you learned how to use cross-validation to evaluate deep neural networks in order to find the model that results in the lowest error rate for a classification problem. You also learned how to improve a given classification model by using cross-validation in order to find the best set of hyperparameters for it. In the next activity, we repeat this activity with a regression task.

ACTIVITY 4.03: MODEL SELECTION USING CROSS-VALIDATION ON A TRAFFIC VOLUME DATASET

In this activity, you are going to practice model selection using cross-validation one more time. Here, we are going to use a simulated dataset that represents a target variable representing the volume of traffic in cars/hour across a city bridge and various normalized features related to traffic data such as the time of day and the traffic volume on the previous day. Our goal is to build a model that predicts the traffic volume across the city bridge given the various features.

The dataset contains **10000** records, and for each of them, **10** attributes/features are included. The goal is to build a deep neural network that receives the **10** features and predicts the traffic volume across the bridge. Since the output is a number, this is a regression problem. Let's get started:

1. Import all the required packages.

2. Print the input and output sizes to check the number of examples in the dataset and the number of features for each example. Also, you can print the range of the output (the output in this dataset represents the median value of owner-occupied homes in thousands of dollars).

3. Define three functions, each returning a different Keras model. The first Keras model will be a shallow neural network with one hidden layer of **size 10** and a **ReLU activation** function. The second Keras model will be a deep neural network with two hidden layers of **size 10** and a **ReLU activation** function in each layer. The third Keras model will be a deep neural network with three hidden layers of **size 10** and a **ReLU activation** function in each layer.

Use the following values as well:

```
optimizer = 'adam', loss = 'mean_squared_error'
```

> **NOTE**
>
> *Steps 4, 5,* and *6* of this activity involve performing **5-fold cross-validation** three, four, and three times, respectively. Therefore, they may take some time to complete.

4. Write the code to loop over the three models and perform **5-fold cross-validation** on each of them (use **epochs=100, batch_size=5**, and **shuffle=False** in this step). Store all the cross-validation scores in a list and print the results. Which model results in the lowest test error rate?

5. Write the code that uses the **epochs = [80, 100]** and **batches = [5, 10]** values for **epochs** and **batch_size**. Perform 5-fold cross-validation for each possible pair on the Keras model that resulted in the lowest test error rate from *step 4*. Store all the cross-validation scores in a list and print the results. Which **epochs** and **batch_size** pair results in the lowest test error rate?

6. Write the code that uses **optimizers = ['rmsprop', 'sgd', 'adam']** and perform **5-fold cross-validation** for each possible optimizer on the Keras model that resulted in the lowest test error rate from *step 4*. Use the **batch_size** and **epochs** values that resulted in the lowest test error rate from *step 5*. Store all the cross-validation scores in a list and print the results. Which **optimizer** results in the lowest test error rate?

After implementing these steps, the expected output will be as follows:

```
optimizer= adam   test error rate =   25.391812739372256
optimizer= sgd   test error rate =   25.140230269432067
optimizer= rmsprop   test error rate =   25.217947859764102
```

> **NOTE**
>
> The solution for this activity can be found on page 391.

In this activity, you learned how to use cross-validation to evaluate deep neural networks in order to find the model that results in the lowest error rate for a **regression problem**. Also, you learned how to improve a given regression model by using cross-validation in order to find the best set of hyperparameters for it.

SUMMARY

In this chapter, you learned about cross-validation, which is one of the most important resampling methods. It results in the best estimation of model performance on independent data. This chapter covered the basics of cross-validation and its two different variations, leave-one-out and k-fold, along with a comparison of them.

Next, we covered the Keras wrapper with scikit-learn, which is a very helpful tool that allows scikit-learn methods and functions that perform cross-validation to be easily applied to Keras models. Following this, you were shown a step-by-step process of implementing cross-validation in order to evaluate Keras deep learning models using the Keras wrapper with scikit-learn.

Finally, you learned that cross-validation estimations of model performance can be used to decide between different models for a particular problem or to decide which parameters (or hyperparameters) should be used for a particular model. You practiced using cross-validation for this purpose by writing user-defined functions in order to perform cross-validation on different models or different possible combinations of hyperparameters and selecting the model or the set of hyperparameters that leads to the lowest test error rate for your final model.

In the next chapter, you will learn that what we did here in order to find the best set of hyperparameters for our model is, in fact, a technique called **hyperparameter tuning** or **hyperparameter optimization**. Also, you will learn how to perform hyperparameter tuning in scikit-learn by using a method called **grid search** and without the need to write user-defined functions to loop over possible combinations of hyperparameters.

5

IMPROVING MODEL ACCURACY

OVERVIEW

This chapter introduces the concept of regularization for neural networks. Regularization aims to prevent the model from overfitting the training data during the training process and provides more accurate results when the model is tested on new unseen data. You will learn to utilize different regularization techniques—L1 and L2 regularization and dropout regularization—to improve model performance. Regularization is an important component as it prevents neural networks from overfitting the training data and helps us build robust, accurate models that perform well on new, unseen data. By the end of this chapter, you will be able to implement a grid search and random search in scikit-learn and find the optimal hyperparameters.

INTRODUCTION

In the previous chapter, we continued to develop our knowledge of creating accurate models with neural networks by experimenting with cross-validation as a method to test how various hyperparameters perform in an unbiased manner. We utilized leave-one-out cross-validation, in which we leave one record out of the training process for use in validation and repeat this for every record in the dataset. Then, we looked at k-fold cross-validation, in which we split the training dataset into **k** folds, train the model on **k-1** folds, and use the final fold for validation. These cross-validation methods allow us to train models with different hyperparameters and test their performance on unbiased data.

Deep learning is not only about building neural networks, training them using an available dataset, and reporting the model accuracy. It involves trying to understand your model and the dataset, as well as moving beyond a basic model by improving it in many aspects. In this chapter, you will learn about two very important groups of techniques for improving machine learning models in general, and deep learning models in particular. These techniques are regularization methods and hyperparameter tuning.

This chapter will further cover regularization methods—specifically, why we need them and how they help. Then, we'll introduce two of the most important and most commonly used regularization techniques. Here, you'll learn about parameter regularization and its two variations, **L1** and **L2** norm regularizations, in great detail. You will then learn about a regularization technique that was specifically designed for neural networks called **dropout regulation**. You will also practice implementing each of these techniques on Keras models by completing activities that involve real-life datasets. We'll end our discussion of regularization by briefly introducing some other regularization techniques that you may find helpful later in your work.

Then, we will talk about the importance of **hyperparameter tuning**, especially for deep learning models, by exploring how tuning the values of hyperparameters can dramatically affect model accuracy, as well as the challenge of tuning the many hyperparameters that require it when building deep neural networks. You will learn two very helpful methods in scikit-learn that you can use for performing hyperparameter tuning on Keras models, the benefits and drawbacks of each method, and how to combine them to gain the most from both. Lastly, you will practice implementing hyperparameter tuning for Keras models using scikit-learn optimizers by completing an activity.

REGULARIZATION

Since deep neural networks are highly flexible models, overfitting is an issue that can often arise when training them. Therefore, one very important part of becoming a deep learning expert is knowing how to detect overfitting, and subsequently how to address the overfitting problem in your model. Overfitting in your models will be clear if your model performs excellently on the training data but performs poorly on new, unseen data.

For example, if you build a model to classify images of dogs and cats into their respective classes and your image classifier performs with high accuracy during the training process but does not perform well on new examples, then this is an indication that your model has overfitted the training data. Regularization techniques are an important group of methods specifically aimed at reducing overfitting in machine learning models.

Understanding regularization techniques thoroughly and being able to apply them to your deep neural networks is an essential step toward building deep neural networks in order to solve real-life problems. In this section, you will learn about the underlying concepts of regularization, which will provide you with the foundation that's required for the following sections, where you will learn how to implement various types of regularization methods using Keras.

THE NEED FOR REGULARIZATION

The main goal of machine learning is to build models that perform well, not only on the examples they are trained on but also on new examples that were not included in the training. A good machine learning model is one that finds the form and the parameters of the true underlying process/function that's producing the training examples but does not capture the noise associated with individual training examples. Such a machine learning model can generalize well to new data that's produced by the same process later.

The approaches we discussed previously—such as splitting a dataset into a training set and a test set, and cross-validation—were all designed to estimate the generalization ability of a trained model. In fact, the term that's used to refer to a test set error and cross-validation error is "generalization error." This simply means the error rate on examples that were not used in training. Once again, the main goal of machine learning is to build models with low generalization error rates.

In *Chapter 3, Deep Learning with Keras*, we discussed two very important issues with machine learning models: overfitting and underfitting. We stated that underfitting is the scenario where the estimated model is not flexible or complex enough to capture all the relations and patterns associated with the true process. This is a model with `high bias` and is detected when the training error is high. On the other hand, overfitting is the scenario where the model that's used for estimating the true process is too flexible or complex. This is a model with `high variance` and is diagnosed when there is a large gap between the training error and the generalization error. An overview of these scenarios for a binary classification problem can be seen in the following images:

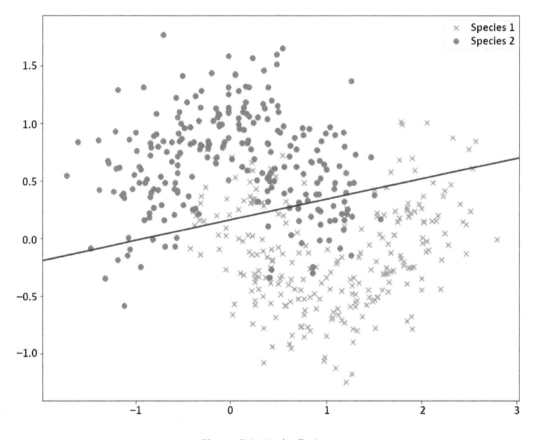

Figure 5.1: Underfitting

As you can see above, Underfitting is a less problematic issue than overfitting. In fact, underfitting can be fixed easily by making the model more flexible/complex. In deep neural networks, this means changing the architecture of the network, making the network larger by adding more layers to it or increasing the number of units in the layers.

Now let's look at overfitting image below:

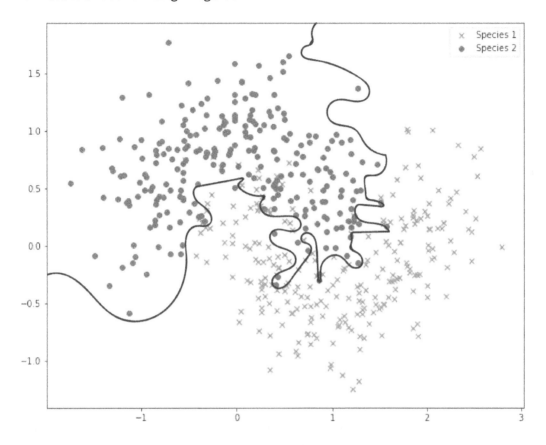

Figure 5.2: Overfitting

Similarly, there are simple solutions for addressing overfitting, such as making the model less flexible/complex (again, by changing the architecture of the network) or providing the network with more training examples. However, making the network less complex sometimes comes at the cost of a dramatic increase in bias or training error rate. The reason for this is that most of the time, the cause of overfitting is not the flexibility of the model but too few training examples. On the other hand, providing more data examples in order to decrease overfitting is not always possible. As a result, finding ways to reduce the generalization error while keeping model complexity and the number of training examples fixed is both important and challenging.

Now let's look at the Right fit image below:

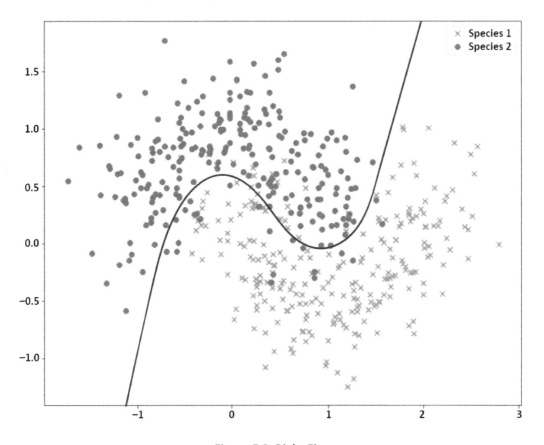

Figure 5.3: Right Fit

That is why we need regularization techniques when building highly flexible machine learning models, such as deep neural networks, to suppress the flexibility of the model so that it cannot overfit to individual examples. In the next section, we will describe how regularization methods reduce the overfitting of models on the training data to reduce the variance in the model.

REDUCING OVERFITTING WITH REGULARIZATION

Regularization methods try to modify the learning algorithm in a way that reduces the variance of the model. By decreasing the variance, regularization techniques intend to reduce the generalization error while not increasing the training error (or, at least, not increasing the training error drastically).

Regularization methods provide some kind of restriction that helps with the stability of the model. There are several ways that this can be achieved. One of the most common ways of performing regularization on deep neural networks is by putting some type of penalizing term on weights to keep the weights small.

Keeping the weights small makes the network less sensitive to noise in individual data examples. Weights in a neural network are, in fact, the coefficients that determine how big or small an effect each processing unit will have on the final output of the network. If the units have large weights, it means that each of them will have a significant influence on the output. Combining all the large influences that are caused by each processing unit will result in many fluctuations in the final output.

On the other hand, keeping the weights small reduces the amount of influence each unit will have on the final output. Indeed, by keeping the weights near zero, some of the units will have almost no effect on the output. Training a large neural network where each unit has little or no effect on the output is the equivalent of training a much simpler network, and so variance and overfitting are reduced. The following figure shows the schematic view of how regularization zeroes out the effect of some units in a large network:

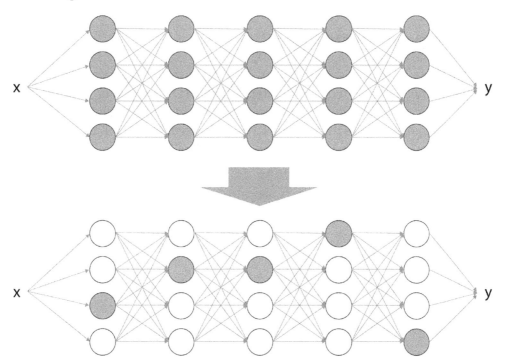

Figure 5.4: Schematic view of how regularization zeroes out the effect of some units in a large network

The preceding diagram is a schematic view of the regularization process. The top network shows a network without regularization, while the bottom network shows an example of a network with regularization in which the white units represent units that have little to no effect on the output because they have been penalized by the regularization process.

So far, we have learned about the concepts behind regularization. In the next section, we will look at the most common methods of regularization for deep learning models—**L1**, **L2**, and dropout regularization—along with how to implement them in Keras.

L1 AND L2 REGULARIZATION

The most common type of regularization for deep learning models is the one that keeps the weights of the network small. This type of regularization is called **weight regularization** and has two different variations: **L2 regularization** and **L1 regularization**. In this section, you will learn about these regularization methods in detail, along with how to implement them in Keras. Additionally, you will practice applying them to real-life problems and observe how they can improve the performance of a model.

L1 AND L2 REGULARIZATION FORMULATION

In weight regularization, a penalizing term is added to the loss function. This term is either the **L2 norm** (the sum of the squared values) of the weights or the **L1 norm** (the sum of the absolute values) of the weights. If the L1 norm is used, then it will be called `L1 regularization`. If the L2 norm is used, then it will be called `L2 regularization`. In each case, the sum is multiplied by a hyperparameter called a **regularization parameter (lambda)**.

Therefore, for `L1 regularization`, the formula is as follows:

*Loss function = Old loss function + lambda * sum of absolute values of the weights*

And for `L2 regularization`, the formula is as follows:

*Loss function = Old loss function + lambda * sum of squared values of the weights*

`Lambda` can take any value between `0` and `1`, where `lambda=0` means no penalty at all (equivalent to a network with no regularization) and `lambda=1` means full penalty.

Like every other hyperparameter, the right value for **lambda** can be selected by trying out different values and observing which value provides a lower generalization error. In fact, it's good practice to start with a network with no regularization and observe the results. Then, you should perform regularization with increasing values of lambda, such as **0.001**, **0.01**, **0.1**, **0.5**, ..., and observe the results in each case in order to figure out how much penalizing on the weight's values is suitable for a particular problem.

In each iteration of the optimization algorithm with regularization, the weights (w) become smaller and smaller. That is why weight regularization is commonly referred to as **weight decay**.

So far, we have only discussed regularizing weights in a deep neural network. However, you need to keep in mind that the same procedure can be applied to biases as well. More precisely, we can update the loss function again by adding a bias penalizing term to it as well and therefore keep the values of biases small during the training of a neural network.

> **NOTE**
>
> If you perform regularization by adding two terms to the loss function (one for penalizing weights and one for penalizing biases), then we call it **parameter regularization** instead of weight regularization.

However, regularizing bias values is not very common in deep learning. The reason for this is that weights are much more important parameters of neural networks. In fact, usually, adding another term to regularize biases will not change the results dramatically in comparison to only regularizing the weight values.

L2 regularization is the most common regularization technique that's used in machine learning in general. The difference between **L1 regularization** and **L2 regularization** is that **L1** results in a sparser weights matrix, meaning there are more weights equal to zero, and therefore more nodes that are completely removed from the network. **L2 regularization**, on the other hand, is more subtle. It decreases the weights drastically, but at the same time leaves you with fewer weights equal to 0. It is also possible to perform both **L1** and **L2 regularization** at the same time.

Now that you have learned about how **L1** and **L2 regularization** work, you are ready to move on to implementing **L1** and **L2 regularization** on deep neural networks in Keras.

L1 AND L2 REGULARIZATION IMPLEMENTATION IN KERAS

Keras provides a regularization API that can be used to add penalizing terms to the loss function in order to regularize weights or biases in each layer of a deep neural network. To define the penalty term or **regularizer**, you need to define the desired regularization method under **keras.regularizers**.

For example, to define an **L1 regularizer** with **lambda=0.01**, you can write this:

```
from keras.regularizers import l1
keras.regularizers.l1(0.01)
```

Similarly, to define an **L2 regularizer** with **lambda=0.01**, you can write this:

```
from keras.regularizers import l2
keras.regularizers.l2(0.01)
```

Finally, to define both **L1** and **L2 regularizers** with **lambda=0.01**, you can write this:

```
from keras.regularizers import l1_l2
keras.regularizers.l1_l2(l1=0.01, l2=0.01)
```

Each of these **regularizers** can be applied to weights and/or biases in a layer. For example, if we would like to apply **L2 regularization** (with **lambda=0.01**) on both the weights and biases of a dense layer with eight nodes, we can write this:

```
from keras.layers import Dense
from keras.regularizers import l2
model.add(Dense(8, kernel_regularizer=l2(0.01), \
        bias_regularizer=l2(0.01)))
```

We will practice implementing **L1** and **L2 regularization** further in *Activity 5.01, Weight Regularization on an Avila Pattern Classifier*, in which you will apply regularization on the deep learning model for the diabetes dataset and observe how the results change in comparison to previous activities.

> **NOTE**
>
> All the activities in this chapter will be developed in Jupyter Notebooks. Please download this book's GitHub repository, along with all the prepared templates, from https://packt.live/2OOBjqq.

ACTIVITY 5.01: WEIGHT REGULARIZATION ON AN AVILA PATTERN CLASSIFIER

The Avila dataset has been extracted from **800** images of the **Avila Bible**, a giant 12th-century Latin copy of the Bible. The dataset consists of various features about the images of the text, such as intercolumnar distance and the margins of the text. The dataset also contains a class label that indicates if a pattern of the image falls into the most frequently occurring category or not. In this activity, you will build a Keras model to perform classification on this dataset according to given network architecture and hyperparameter values. The goal is to apply different types of weight regularization on the model and observe how each type changes the result.

In this activity, we will use the **training set/test set** approach to perform the evaluation for two reasons. First, since we are going to try several different regularizers, performing cross-validation will take a long time. Second, we would like to plot the trends in the training error and the test error in order to understand, in a visual way, how regularization prevents the model from overfitting to data examples.

Follow these steps to complete this activity:

1. Load the dataset from the **data** subfolder of **Chapter05** from GitHub using **X = pd.read_csv('../data/avila-tr_feats.csv')** and **y = pd.read_csv('../data/avila-tr_target.csv')**. Split the dataset into a training set and a test set using the **sklearn.model_selection. train_test_split** method. Hold back **20%** of the data examples for the test set.

2. Define a Keras model with three hidden layers, the first of **size** **10**, the second of **size** **6**, and the third of **size** **4**, to perform the classification. Use these values for the hyperparameters: **activation='relu'**, **loss='binary_crossentropy'**, **optimizer='sgd'**, **metrics=['accuracy']**, **batch_size=20**, **epochs=100**, and **shuffle=False**.

3. Train the model on the training set and evaluate it with the test set. Store the training loss and test loss at every iteration. After training is complete, plot the trends in **training error** and **test error** (change the limits of the vertical axis to (**0, 1**) so that you can observe the changes in losses better). What is the minimum error rate on the test set?

4. Add **L2 regularizers** with **lambda=0.01** to the hidden layers of your model and repeat the training. After training is complete, plot the trends in training error and test error. What is the minimum error rate on the test set?

5. Repeat the previous step for **lambda=0.1** and **lambda=0.005**, train the model for each value of **lambda**, and report the results. Which value of **lambda** is a better choice for performing **L2 regularization** on this deep learning model and this dataset?

6. Repeat the previous step, this time with **L1 regularizers** for **lambda=0.01** and **lambda=0.005**, train the model for each value of **lambda**, and report the results. Which value of **lambda** is a better choice for performing **L1 regularization** on this deep learning model and this dataset?

7. Add **L1_L2 regularizers** with the **L1 lambda=0.005** and the **L2 lambda=0.005** to the hidden layers of your model and repeat the training. After training is complete, plot the trends in training error and test error. What is the minimum error rate on the test set?

After implementing these steps, you should get the following expected output:

```
Best Accuracy on Validation Set = 0.5925215482711792
```

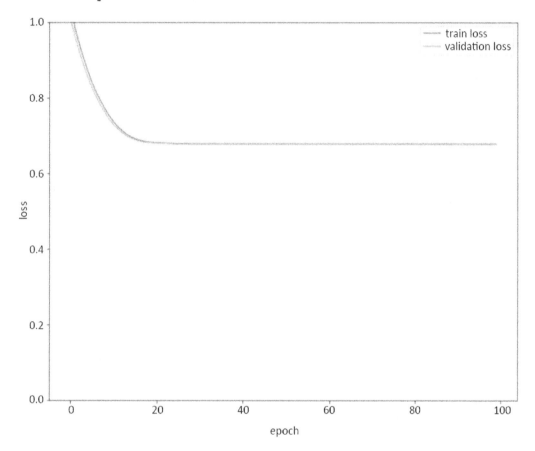

Figure 5.5: A plot of the training error and validation error during training for the model
with L1 lambda equal to 0.005 and L2 lambda equal to 0.005

> **NOTE**
>
> The solution for this activity can be found on page 398.

In this activity, you practiced implementing **L1** and **L2 weight
regularizations** for a real-life problem and compared the results of the
regularized model with those of a model without any regularization. In the next
section, we will explore the regulation of a different technique, known as
dropout regularization.

DROPOUT REGULARIZATION

In this section, you will learn how dropout regularization works, how it helps with reducing overfitting, and how to implement it using Keras. Lastly, you will practice what you have learned about dropout by completing an activity involving a real-life dataset.

PRINCIPLES OF DROPOUT REGULARIZATION

Dropout regularization works by randomly removing nodes from a neural network during training. More precisely, dropout sets up a probability on each node. This probability refers to the chance that the node is included in the training at each iteration of the learning algorithm. Imagine we have a large neural network where a dropout chance of **0.5** is assigned to each node. In such a case, at each iteration, the learning algorithm flips a coin for each node to decide whether that node will be removed from the network or not. An illustration of such a process can be seen in the following diagram:

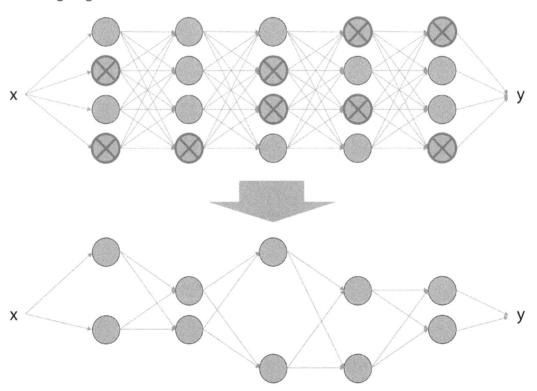

Figure 5.6: Illustration of removing nodes from a deep neural network using dropout regularization

This process is repeated at each iteration; this means that, at each iteration, randomly selected nodes are removed from the network, which means the parameter-updating process is done on a different smaller network. For example, the network shown at the bottom of the preceding diagram would be used for one iteration of the training only. For the next iteration, some other randomly selected nodes would be crossed out from the top network so the network that results from removing those nodes would be different from the bottom network in the diagram.

When some nodes are chosen to be removed/ignored in an iteration of a learning algorithm, it means that they won't participate in the parameter-updating process at all in that iteration. More precisely, the forward propagation to predict the output, the loss computation, and the backpropagation to compute the derivatives are all to be done on the smaller network with some nodes removed. Consequently, parameter updating will only be done on the nodes that are present in the network in that iteration; the weights and biases of removed nodes won't be updated.

However, it is important to keep in mind that to evaluate the performance of the model on the test set or hold-out set, the original complete network is always used. If we perform the evaluation of a network with random nodes deleted from it, the noise will be introduced to the results, and this is not desirable.

In **dropout regularization**, **training** is always performed on the networks that result from randomly selected nodes being removed from the original network. **Evaluation** is always performed using the original network. In the next section, we will gain an understanding of why dropout regularization helps prevent overfitting.

REDUCING OVERFITTING WITH DROPOUT

In this section, we are going to discuss the concepts behind dropout as a regularization method. As we discussed previously, the goal of regularization techniques is to prevent a model from overfitting data. Therefore, we are going to look at how randomly removing a portion of nodes from a neural network helps reduce variance and overfitting.

The most obvious explanation of why removing random nodes from the network prevents overfitting is that by removing nodes from a network, we are performing training on a smaller network in comparison to the original network. As you learned previously, a smaller neural network provides less flexibility, so the chance of the network overfitting to data is lower.

There is another reason why dropout regularization does such a good job of reducing overfitting. By randomly removing inputs at each layer in a deep neural network, the overall network becomes less sensitive to single inputs. We know that, while training a neural network, the weights are updated in a way that the final model will fit to the training examples. By removing some of the weights from the training process at random, dropout forces other weights to participate in learning the patterns related to the training examples at that iteration, and so the final weight values will better spread out more.

In other words, instead of some weights updating too much in order to fit some input values, all the weights learn to participate in learning those input values and, consequently, overfitting decreases. This is why performing dropout results in a much more robust model—performing better on new, unseen data—in comparison to simply using a smaller network. In fact, **dropout regularization** tends to work better on larger networks.

Now that you have learned all about the underlying procedure of dropout and the reasons behind its effectiveness, we can move on to implementing **dropout regularization** in Keras.

EXERCISE 5.01: DROPOUT IMPLEMENTATION IN KERAS

Dropout regularization is provided as a core layer in Keras. As such, you can add dropout to your model in the same way that you would add layers to your network. When defining a dropout layer in Keras, you need to provide the **rate** hyperparameter as an argument. **rate** can take any value between **0** and **1** and determines the portions of the input units to be removed or ignored. In this exercise, you will learn the step-by-step process of implementing a Keras deep learning model with dropout layers.

Our simulated dataset represents various measurements of trees, such as height, the number of branches, and the girth of the trunk at the base. Our goal is to classify the records into either deciduous (a class value of **1**) or coniferous (a class value of **0**) type trees based on the measurements given. The dataset consists of **10000** records that consist of two classes, representing the two tree types, and each data example has **10** feature values. Follow these steps to complete this exercise:

1. First, execute the following code block to load in the dataset and split the dataset into a **training set** and a **test set**:

```
# Load the data
import pandas as pd
X = pd.read_csv('../data/tree_class_feats.csv')
y = pd.read_csv('../data/tree_class_target.csv')


"""
Split the dataset into training set and test set with a 80-20 ratio
"""
from sklearn.model_selection import train_test_split
seed = 1
X_train, X_test, \
y_train, y_test = train_test_split(X, y, \
                                   test_size=0.2, \
                                   random_state=seed)
```

2. Import all the necessary dependencies. Build a four-layer Keras sequential model without **dropout regularization**. Build the network with 16 units in the first hidden layer, **12** units in the second hidden layer, **8** units in the third hidden layer, and **4** units in the fourth hidden layer, all with **ReLU activation** functions. Add an output layer with a **sigmoid activation** function:

```
#Define your model
from keras.models import Sequential
from keras.layers import Dense, Activation
import numpy as np
from tensorflow import random

np.random.seed(seed)
random.set_seed(seed)

model_1 = Sequential()
model_1.add(Dense(16, activation='relu', input_dim=10))
model_1.add(Dense(12, activation='relu'))
model_1.add(Dense(8, activation='relu'))
model_1.add(Dense(4, activation='relu'))
model_1.add(Dense(1, activation='sigmoid'))
```

3. Compile the model with **binary cross-entropy** as the **loss** function and **sgd** as the optimizer and train the model for **300** epochs with **batch_size=50** on the **training set**. Then, evaluate the trained model on the **test set**:

```
model_1.compile(optimizer='sgd', loss='binary_crossentropy')
# train the model
model_1.fit(X_train, y_train, epochs=300, batch_size=50, \
            verbose=0, shuffle=False)
# evaluate on test set
print("Test Loss =", model_1.evaluate(X_test, y_test))
```

Here's the expected output:

```
2000/2000 [==============================] - 0s 23us/step
Test Loss = 0.1697693831920624
```

Therefore, the test error rate for predicting the species of tree after training the model for **300** epochs is equal to **16.98%**.

4. Redefine the model with the same number of layers and same size in each layer as the prior model. However, add a **dropout regularization** of **rate=0.1** to the first hidden layer of your model and repeat the compilation, training, and evaluation steps of the model on the test data:

```
"""
define the keras model with dropout in the first hidden layer
"""
from keras.layers import Dropout
np.random.seed(seed)
random.set_seed(seed)
model_2 = Sequential()
model_2.add(Dense(16, activation='relu', input_dim=10))
model_2.add(Dropout(0.1))
model_2.add(Dense(12, activation='relu'))
model_2.add(Dense(8, activation='relu'))
model_2.add(Dense(4, activation='relu'))
model_2.add(Dense(1, activation='sigmoid'))

model_2.compile(optimizer='sgd', loss='binary_crossentropy')
# train the model
```

```
model_2.fit(X_train, y_train, \
            epochs=300, batch_size=50, \
            verbose=0, shuffle=False)
# evaluate on test set
print("Test Loss =", model_2.evaluate(X_test, y_test))
```

Here's the expected output:

```
2000/2000 [==============================] - 0s 29us/step
Test Loss = 0.16891103076934816
```

After adding a dropout regularization of **rate=0.1** to the first layer of the network, the test error rate is reduced from **16.98%** to **16.89%**.

5. Redefine the model with the same number of layers and the same size in each layer as the prior model. However, add a dropout regularization of **rate=0.2** to the first hidden layer and **rate=0.1** to the remaining layers of your model and repeat the compilation, training, and evaluation steps of the model on the test data:

```
# define the keras model with dropout in all hidden layers
np.random.seed(seed)
random.set_seed(seed)

model_3 = Sequential()
model_3.add(Dense(16, activation='relu', input_dim=10))
model_3.add(Dropout(0.2))
model_3.add(Dense(12, activation='relu'))
model_3.add(Dropout(0.1))
model_3.add(Dense(8, activation='relu'))
model_3.add(Dropout(0.1))
model_3.add(Dense(4, activation='relu'))
model_3.add(Dropout(0.1))
model_3.add(Dense(1, activation='sigmoid'))

model_3.compile(optimizer='sgd', loss='binary_crossentropy')
# train the model
model_3.fit(X_train, y_train, epochs=300, \
            batch_size=50, verbose=0, shuffle=False)
# evaluate on test set
print("Test Loss =", model_3.evaluate(X_test, y_test))
```

Here's the expected output:

```
2000/2000 [==============================] - 0s 40us/step
Test Loss = 0.19390961921215058
```

By keeping the dropout regularization of **rate=0.2** in the first layer while adding dropout regularizations of **rate=0.1** to the subsequent layers, the test error rate increased from **16.89%** to **19.39%**. Like the **L1** and **L2 regularizations**, adding too much dropout can prevent the model from learning the underlying function associated with the training data and leads to higher bias than without dropout regularization.

As you saw in this exercise, you can also apply dropout with different rates to the different layers depending on how much overfitting you think can happen in those layers. Usually, we prefer not to perform dropout on the input layer and the output layer. Regarding the hidden layers, we need to tune the **rate** values and observe the results in order to decide what value is best suited to a particular problem.

> **NOTE**
>
> To access the source code for this specific section, please refer to https://packt.live/3iugM7K.
>
> You can also run this example online at https://packt.live/31HISYo.

In the following activity, you will practice implementing deep learning models along with dropout regularization in Keras on the Traffic Volume dataset.

ACTIVITY 5.02: DROPOUT REGULARIZATION ON THE TRAFFIC VOLUME DATASET

In *Activity 4.03, Model Selection Using Cross-Validation on a Traffic Volume Dataset*, of *Chapter 4, Evaluating Your Model with Cross-Validation Using Keras Wrappers*, you used the Traffic Volume dataset to build a model for predicting the volume of traffic across a city bridge when given various normalized features related to traffic data such as the time of day and the volume on the previous day, among others. The dataset contains **10000** records and for each of them, **10** attributes/features are included in the dataset.

In this activity, you will start with the model from *Activity 4.03, Model Selection Using Cross-Validation on a Traffic Volume Dataset*, of *Chapter 4, Evaluating Your Model with Cross-Validation Using Keras Wrappers*. You will use the training set/test set approach to train and evaluate the model, plot the trends in training error and the generalization error, and observe the model overfitting data examples. Then, you will attempt to improve model performance by addressing the overfitting issue through the use of dropout regularization. In particular, you will try to find out which layers you should add dropout regularization to and what **rate** value will improve this specific model the most. Follow these steps to complete this activity:

1. Load the dataset using the pandas **read_csv** function. The dataset is also stored in the **data** subfolder of the *Chapter05* GitHub repository. Split the dataset into a training set and a test set with an **80–20** ratio.

2. Define a Keras model with two hidden layers of **size 10** to predict the traffic volume. Use these values for the following hyperparameters: **activation='relu', loss='mean_squared_error', optimizer='rmsprop', batch_size=50, epochs=200**, and **shuffle=False**.

3. Train the model on the training set and evaluate on the test set. Store the training loss and test loss at every iteration.

4. After training is completed, plot the trends in training error and test error. What are the lowest error rates on the training set and the test set?

5. Add dropout regularization with **rate=0.1** to the first hidden layer of your model and repeat the training process (since training with dropout takes longer, train for **200** epochs). After training is completed, plot the trends in training error and test error. What are the lowest error rates on the training set and the test set?

6. Repeat the previous step, this time adding dropout regularization with **rate=0.1** to both hidden layers of your model and train the model and report the results.

7. Repeat the previous step, this time with **rate=0.2** on the first layer and **0.1** on the second layer and train the model and report the results.

8. Which dropout regularization has resulted in the best performance on this deep learning model and this dataset so far?

After implementing these steps, you should get the following expected output:

```
Lowest error on training set =  802.1680335998535
Lowest error on validation set =  115.19878396987914
```

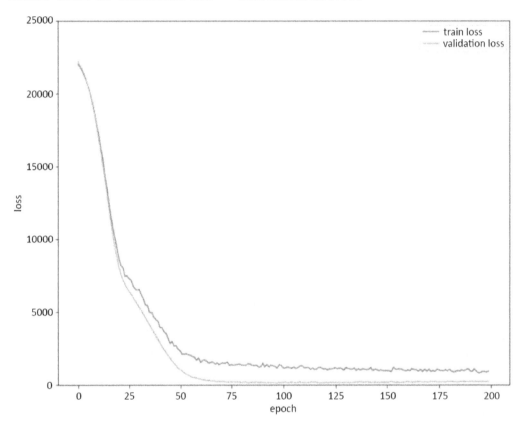

Figure 5.7: A plot of training errors and validation errors while training the model with dropout regularization, with rate=0.2 in the first layer and rate=0.1 in the second layer

NOTE

The solution for this activity can be found on page 413.

In this activity, you learned how to implement dropout regularization in Keras and practiced using it on a problem involving the Traffic Volume dataset. **Dropout regularization** is specifically designed for the purpose of reducing overfitting in neural networks and works by randomly removing nodes from a neural network during the training process. This procedure results in a neural network with well spread out weight values, which leads to less overfitting in individual data examples. In the next section, we will discuss other regularization methods that can be applied to prevent a model overfitting the training data.

OTHER REGULARIZATION METHODS

In this section, you will briefly learn about some other regularization techniques that are commonly used and have been shown to be effective in deep learning. It is important to keep in mind that regularization is a wide-ranging and active research field in machine learning. As a result, covering all the available regularization methods in one chapter is not possible (and most likely not necessary, especially in a book on applied deep learning). Therefore, in this section, we will briefly cover three more regularization methods, called **early stopping**, **data augmentation**, and **adding noise**. You will learn about their underlying ideas and gain a few tips and recommendations on how to use them.

EARLY STOPPING

Earlier in this chapter, we discussed that the main assumption in machine learning is that there is a true function or process that produces training examples. However, this process is unknown and there is no explicit way to find it. Not only is there no way to find the exact underlying process but choosing a model with the right level of flexibility or complexity for estimating the process is challenging as well. Therefore, one good practice is to select a highly flexible model, such as a deep neural network, to model the process and monitor the training process carefully.

By monitoring the training process, we can train the model just enough for it to capture the form of the process, and we can stop the training right before it starts to overfit to individual data examples. This is the underlying idea behind early stopping. We discussed the idea of early stopping briefly in the *Model Evaluation* section of *Chapter 3*, *Deep Learning with Keras*. We stated that, by monitoring and observing the changes in **training error** and **test error** during training, we can determine how little training is too little and how much training is too much.

The following plot shows a view of the changes in training error and test error when a highly flexible model is trained on a dataset. As we can see, the training needs to stop in the region labeled **Right Fit** to avoid overfitting:

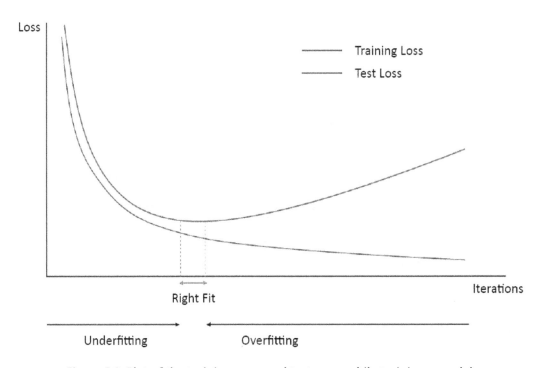

Figure 5.8: Plot of the training error and test error while training a model

In *Chapter 3, Deep Learning with Keras*, we practiced storing and plotting changes in training error and test error in order to identify overfitting. You learned that you can provide a validation set or test set when training a Keras model and store the metrics values for each of them at each epoch of training by using the following code:

```
history=model.fit(X_train, y_train, validation_data=(X_test, y_test), \
                epochs=epochs)
```

In this section, you are going to learn how to implement early stopping in Keras. This means forcing the Keras model to stop the training when a desired metric—for example, the **test error rate**—is not improving anymore. In order to do so, you need to define an **EarlyStopping()** callback and provide it as an argument to **model.fit()**.

When defining an **EarlyStopping()** callback, you need to provide it with the right arguments. The first argument is **monitor**, which determines what metric will be monitored during training for the purpose of performing early stopping. Usually, **monitor='val_loss'** is a good choice, meaning that we would like to monitor the test error rate.

Also, depending on what argument you have chosen for the **monitor**, you need to set the **mode** argument to either **'min'** or **'max'**. If the metric is error/loss, we would like to minimize it. For example, the following code block defines an **EarlyStopping()** callback that monitors the test error during training and detects if it is not decreasing anymore:

```
from keras.callbacks import EarlyStopping
es_callback = EarlyStopping(monitor='val_loss', mode='min')
```

If there are a lot of fluctuations or noise in the error rates, it is probably not a good idea to stop the training when the loss begins to increase at all. For this reason, we can set the **patience** argument to a number of epochs to give the early stopping method some time to monitor the desired metric for longer before stopping the training process:

```
es_callback = EarlyStopping(monitor='val_loss', \
                            mode='min', patience=20)
```

We can also modify the **EarlyStopping()** callback to stop the training process if a minimal improvement in the **monitor** metric has not happened in the past **epoch**, or the **monitor** metric has reached a baseline level:

```
es_callback = EarlyStopping(monitor='val_loss', \
                            mode='min', min_delta=1)
es_callback = EarlyStopping(monitor='val_loss', \
                            mode='min', baseline=0.2)
```

After defining the **EarlyStopping()** callback, you can provide it as a **callbacks** argument to **model.fit()** and train the model. The training will automatically stop according to the **EarlyStopping()** callback:

```
history=model.fit(X_train, y_train, validation_data=(X_test, y_test), \
                  epochs=epochs, callbacks=[es_callback])
```

We will explore how early stopping can be achieved in practice in the next exercise.

EXERCISE 5.02: IMPLEMENTING EARLY STOPPING IN KERAS

In this exercise, you will learn how to implement early stopping on a Keras deep learning model. The dataset we will use is a simulated dataset that represents various measurements of trees, such as height, the number of branches and the girth of the trunk at the base. Our goal is to classify the records into either deciduous or coniferous trees based on the measurements given.

First, execute the following code block to load a simulated dataset of **10000** records that consist of two classes representing the two tree species, with a class value of **1** for deciduous tree species and a class value of **0** for coniferous tree species. Each record has **10** feature values.

The goal is to build a model in order to predict the species of the tree when given the measurements of the tree. Now, let's go through the steps:

1. Load the dataset using the pandas **read_csv** function and split the dataset in an **80-20** split using the **train_test_split** function:

```
# Load the data
import pandas as pd
X = pd.read_csv('../data/tree_class_feats.csv')
y = pd.read_csv('../data/tree_class_target.csv')


"""
Split the dataset into training set and test set with an 80-20 ratio
"""
from sklearn.model_selection import train_test_split
seed=1
X_train, X_test, \
y_train, y_test = train_test_split(X, y, test_size=0.2, \
                                   random_state=seed)
```

2. Import all the necessary dependencies. Build a three-layer Keras sequential model without early stopping. The first layer will have **16** units, the second layer will have **8** units, and the third layer will have **4** units, all with **ReLU activation** functions. Add the **output layer** with a **sigmoid activation function**:

```
# Define your model
from keras.models import Sequential
from keras.layers import Dense, Activation
import numpy as np
```

```
from tensorflow import random

np.random.seed(seed)
random.set_seed(seed)

model_1 = Sequential()
model_1.add(Dense(16, activation='relu', \
                  input_dim=X_train.shape[1]))
model_1.add(Dense(8, activation='relu'))
model_1.add(Dense(4, activation='relu'))
model_1.add(Dense(1, activation='sigmoid'))
```

3. Compile the model with the **loss** function as binary cross-entropy and the optimizer as **SGD**. Train the model for **300** epochs with **batch_size=50**, all while storing the **training error** and the **test error** at every iteration:

```
model_1.compile(optimizer='sgd', loss='binary_crossentropy')
# train the model
history = model_1.fit(X_train, y_train, \
                      validation_data=(X_test, y_test), \
                      epochs=300, batch_size=50, \
                      verbose=0, shuffle=False)
```

4. Import the required packages for plotting:

```
import matplotlib.pyplot as plt
import matplotlib
%matplotlib inline
```

5. Plot the **training error** and **test error** that are stored in the variable that was created during the fitting process:

```
matplotlib.rcParams['figure.figsize'] = (10.0, 8.0)
plt.plot(history.history['loss'])
plt.plot(history.history['val_loss'])
plt.ylim(0,1)
plt.ylabel('loss')
plt.xlabel('epoch')
plt.legend(['train loss', 'validation loss'], \
           loc='upper right')
```

Here's the expected output:

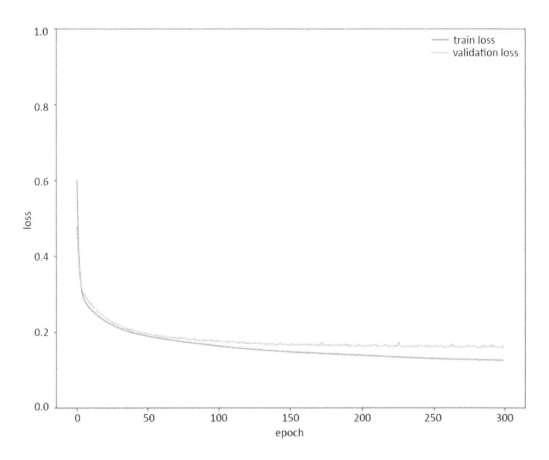

Figure 5.9: Plot of the training error and validation error while training the model without early stopping

As you can see from the preceding plot, training the model for **300** epochs results in a gap that grows between the **training error** and **validation error**, which is indicative of overfitting beginning to happen.

6. Redefine the model by creating the model with the same number of layers and with the same number of units within each layer. This ensures the model is initialized in the same way. Add a callback **es_callback = EarlyStopping(monitor='val_loss', mode='min')** to the training process. Repeat *step 4* to plot the training error and validation error:

```
#Define your model with early stopping on test error
from keras.callbacks import EarlyStopping
np.random.seed(seed)
random.set_seed(seed)


model_2 = Sequential()
model_2.add(Dense(16, activation='relu', \
                  input_dim=X_train.shape[1]))
model_2.add(Dense(8, activation='relu'))
model_2.add(Dense(4, activation='relu'))
model_2.add(Dense(1, activation='sigmoid'))
"""
Choose the loss function to be binary cross entropy and the optimizer
to be SGD for training the model
"""
model_2.compile(optimizer='sgd', loss='binary_crossentropy')
# define the early stopping callback
es_callback = EarlyStopping(monitor='val_loss', \
                            mode='min')
# train the model
history=model_2.fit(X_train, y_train, \
                    validation_data=(X_test, y_test), \
                    epochs=300, batch_size=50, \
                    callbacks=[es_callback], verbose=0, \
                    shuffle=False)
```

7. Now plot the loss values:

```
# plot training error and test error
matplotlib.rcParams['figure.figsize'] = (10.0, 8.0)
plt.plot(history.history['loss'])
plt.plot(history.history['val_loss'])
plt.ylim(0,1)
plt.ylabel('loss')
plt.xlabel('epoch')
plt.legend(['train loss', 'validation loss'], \
            loc='upper right')
```

Here's the expected output:

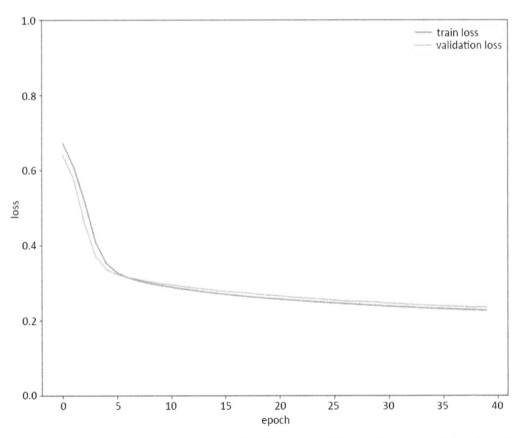

Figure 5.10: Plot of training error and validation error while training the model with early stopping (patience=0)

By adding the early stopping callback with **patience=0** to the model, the training process automatically stops after about **39** epochs.

8. Repeat *step 5* while adding **patience=10** to your early stopping callback. Repeat *step 3* to plot the **training error** and **validation error**:

```
"""
Define your model with early stopping on test error with patience=10
"""

from keras.callbacks import EarlyStopping
np.random.seed(seed)
random.set_seed(seed)

model_3 = Sequential()
model_3.add(Dense(16, activation='relu', \
                  input_dim=X_train.shape[1]))
model_3.add(Dense(8, activation='relu'))
model_3.add(Dense(4, activation='relu'))
model_3.add(Dense(1, activation='sigmoid'))
"""
Choose the loss function to be binary cross entropy and the optimizer
to be SGD for training the model
"""
model_3.compile(optimizer='sgd', loss='binary_crossentropy')
# define the early stopping callback
es_callback = EarlyStopping(monitor='val_loss', \
                            mode='min', patience=10)
# train the model
history=model_3.fit(X_train, y_train, \
                    validation_data=(X_test, y_test), \
                    epochs=300, batch_size=50, \
                    callbacks=[es_callback], verbose=0, \
                    shuffle=False)
```

9. Then plot the loss again:

```
# plot training error and test error
matplotlib.rcParams['figure.figsize'] = (10.0, 8.0)
plt.plot(history.history['loss'])
plt.plot(history.history['val_loss'])
plt.ylim(0,1)
plt.ylabel('loss')
plt.xlabel('epoch')
plt.legend(['train loss', 'validation loss'], \
           loc='upper right')
```

Here's the expected output:

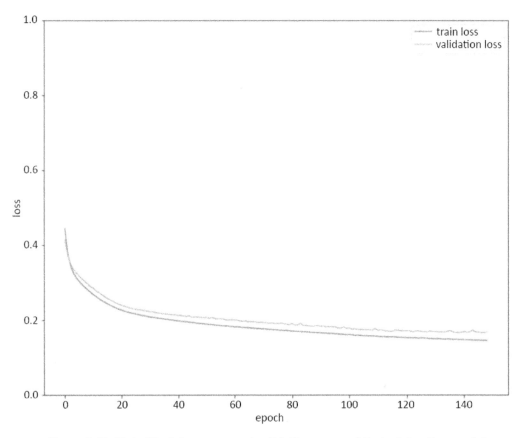

Figure 5.11: Plot of training error and validation error while training the model with early stopping (patience=10)

By adding the early stopping callback with **patience=10** to the model, the training process automatically stops after about **150** epochs.

In this exercise, you learned how to stop the model to prevent your Keras model from overfitting the training data. To do this, you utilized the **EarlyStopping** callback and trained the model with it. We used this callback to stop the model any time the validation loss increased and added a **patience** parameter, which waits for a given number of epochs before stopping. We practiced using this callback on a problem involving the Traffic Volume dataset to train our Keras model.

> **NOTE**
>
> To access the source code for this specific section, please refer to https://packt.live/3iuM4eL.
>
> You can also run this example online at https://packt.live/38AbweB.

In the next section, we will discuss other regularization methods that can be applied to prevent overfitting.

DATA AUGMENTATION

Data augmentation is a regularization technique that tries to address overfitting by training the model on more training examples in an inexpensive way. In data augmentation, the available data is transformed in different ways and fed to the model as new training data. This type of regularization has been shown to be effective, especially for some specific applications, such as object detection/recognition in computer vision and speech processing.

For example, in computer vision applications, you can simply double or triple the size of your training dataset by adding mirrored versions and rotated versions of each image to the dataset. The new training examples that are generated by these transformations are obviously not as good as the original training examples. However, they are shown to improve the model in terms of overfitting.

One challenging aspect of performing data augmentation is choosing the right transformations to be performed on data. Transformations need to be selected carefully, depending on the type of dataset and the application.

ADDING NOISE

The underlying idea behind regularizing a model by adding noise to the data is the same as that for data augmentation regularization. Training a deep neural network on a small dataset increases the chance of the network memorizing single data examples as opposed to capturing the relations between inputs and outputs.

This will result in poor performance on new data later, which is indicative of the model overfitting the training data. In contrast, training a model on a large dataset increases the chance of the model capturing the true underlying process instead of memorizing single data points, and therefore reduces the chances of overfitting.

One way to expand the training data and reduce overfitting is to generate new data examples by injecting noise into the available data. This type of regularization has been shown to reduce overfitting to an extent that is comparable to weight regularization techniques.

By adding different versions of a single example to the training data (each created by adding a small amount of noise to the original example), we can ensure that the model will not fit the noise in the data. Additionally, increasing the size of the training dataset by including these modified examples provides the model with a better representation of the underlying data generation process and increases the chance of the model learning the true process.

In deep learning applications, you can improve model performance by adding noise to the weights or activations of the hidden layers, or gradients of the network, or even to the output layer, as well as by adding noise to the training examples (input layer). Deciding where to add noise in a deep neural network is another challenge that needs to be addressed by trying different networks and observing the results.

In Keras, you can easily define noise as a layer and add it to your model. For example, to add **Gaussian noise** with a **standard deviation** of **0.1** (the mean is equal to **0**) to your model, you can write this:

```
from keras.layers import GaussianNoise
model.add(GaussianNoise(0.1))
```

The following code will add **Gaussian noise** to the outputs/activations of the first hidden layer of the model:

```
model = Sequential()
model.add(Dense(4, input_dim=30, activation='relu'))
model.add(GaussianNoise(0.01))
model.add(Dense(4, activation='relu'))
```

```
model.add(Dense(4, activation='relu'))
model.add(Dense(1, activation='sigmoid'))
```

In this section, you learned about three regularization methods: **early stopping**, **data augmentation**, and **adding noise**. In addition to their basic concepts and procedures, you also learned about how they reduce overfitting and were given some tips and recommendations on how to use them. In the next section, you will learn how to tune hyperparameters using functions provided by scikit-learn. By doing this, we can incorporate Keras models into a scikit-learn workflow.

HYPERPARAMETER TUNING WITH SCIKIT-LEARN

Hyperparameter tuning is a very important technique for improving the performance of deep learning models. In *Chapter 4, Evaluating Your Model with Cross-Validation Using Keras Wrappers*, you learned about using a Keras wrapper with scikit-learn, which allows for Keras models to be used in a scikit-learn workflow. As a result, different general machine learning and data analysis tools and methods that are available in scikit-learn can be applied to Keras deep learning models. Among those methods are scikit-learn hyperparameter optimizers.

In the previous chapter, you learned how to perform hyperparameter tuning by writing user-defined functions to loop over possible values for each hyperparameter. In this section, you will learn how to perform it in a much easier way by using the various hyperparameter optimization methods that are available in scikit-learn. You will also get to practice applying those methods by completing an activity involving a real-life dataset.

GRID SEARCH WITH SCIKIT-LEARN

So far, we have established that building deep neural networks involves making decisions about several hyperparameters. The list of hyperparameters includes the number of hidden layers, the number of units in each hidden layer, the activation function for each layer, the loss function for the network, the type of optimizer and its parameters, the type of regularizer and its parameters, the batch size, the number of epochs, and others. We also observed that different values of hyperparameters can affect the performance of a model significantly.

Therefore, finding the best values for hyperparameters is one of the most important and challenging parts of becoming a deep learning expert. Since there are no absolute rules for picking the hyperparameters that work for every dataset and every problem, deciding on the values of hyperparameters needs to be done through trial and error for each particular problem. This process of training and evaluating models with different hyperparameters and deciding about the final hyperparameters based on model performance is called **hyperparameter tuning** or **hyperparameter optimization**.

Having a range or a set of possible values for each hyperparameter that we are interested in tuning can create a grid such as the one shown in the following image. Therefore, hyperparameter tuning can be seen as a grid search problem; we would like to try every cell in the grid (every possible combination of hyperparameters) and find the one cell that results in the best performance for the model:

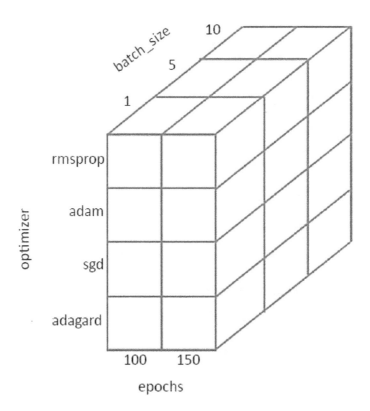

Figure 5.12: A hyperparameter grid created by some values for optimizer, batch_size, and epochs

Scikit-learn provides a parameter optimizer called **GridSearchCV()** to perform this exhaustive grid search. **GridSearchCV()** receives the model as the **estimator** argument and the dictionary containing all possible values for the hyperparameters as the **param_grid** argument. Then, it goes through every point in the grid, performs cross-validation on the model using the hyperparameter values at that point, and returns the best cross-validation score, along with the values of the hyperparameters that led to that score.

In the previous chapter, you learned that in order to use Keras models in scikit-learn, you need to define a function that returns a Keras model. For example, the following code block defines a Keras model that we would like to perform hyperparameter tuning on later:

```
from keras.models import Sequential
from keras.layers import Dense
def build_model():
    model = Sequential(optimizer)
    model.add(Dense(10, input_dim=X_train.shape[1], \
                    activation='relu'))
    model.add(Dense(10, activation='relu'))
    model.add(Dense(1))
    model.compile(loss='mean_squared_error', \
             optimizer= optimizer)
    return model
```

The next step would be to define the grid of parameters. For example, say we would like to tune over **optimizer=['rmsprop', 'adam', 'sgd', 'adagrad'],
epochs = [100, 150], batch_size = [1, 5, 10]**. To do so, we would write the following:

```
optimizer = ['rmsprop', 'adam', 'sgd', 'adagrad']
epochs = [100, 150]
batch_size = [1, 5, 10]

param_grid = dict(optimizer=optimizer, epochs=epochs, \
                batch_size= batch_size)
```

Now that the hyperparameter grid has been created, we can create the wrapper so that we can build the interface for the Keras model and use it as an estimator to perform the grid search:

```
from keras.wrappers.scikit_learn import KerasRegressor
model = KerasRegressor(build_fn=build_model, \
                       verbose=0, shuffle=False)


from sklearn.model_selection import GridSearchCV
grid_search = GridSearchCV(estimator=model, \
                           param_grid=param_grid, cv=10)
results = grid_search.fit(X, y)
```

The preceding code through goes through every cell in the grid exhaustively and performs 10-fold cross-validation using hyperparameter values in each cell (here, it performs **10-fold cross-validation** 4*2*3=24 times). Then, it returns the cross-validation score for each of these **24** cells, along with the one that resulted in the best score.

> **NOTE**
>
> Performing k-fold cross-validation on many possible combinations of hyperparameters sure takes a long time. For this reason, you can parallelize the process by passing the **n_jobs=-1** argument to **GridSearchCV()**, which results in using every processor available to perform the grid search. The default value for this argument is **n_jobs=1**, which means no parallelization.

Creating a hyperparameter grid is just one way to iterate through hyperparameters to find the optimal selection. Another way is to simply randomize the selection of hyperparameters, which we will learn about in the next topic.

RANDOMIZED SEARCH WITH SCIKIT-LEARN

As you may have realized, an exhaustive grid search may not be the best choice for tuning the hyperparameters of a deep learning model since it is not very efficient. There are many hyperparameters in deep learning, and especially if you would like to try a large range of values for each, an exhaustive grid search would simply take too long to complete. An alternative way to perform hyperparameter optimization is to perform random sampling on the grid and perform k-fold cross-validation on some randomly selected cells. Scikit-learn provides an optimizer called **RandomizedSearchCV()** to perform a random search for the purpose of hyperparameter optimization.

For example, we can change the code from the previous section from an exhaustive grid search to a random search like so:

```
from keras.wrappers.scikit_learn import KerasRegressor
model = KerasRegressor(build_fn=build_model, verbose=0)


from sklearn.model_selection import RandomizedSearchCV
grid_search = RandomizedSearchCV(estimator=model, \
                                 param_distributions=param_grid, \
                                 cv=10, n_iter=12)
results = grid_search.fit(X, y)
```

Notice that **RandomizedSearchCV()** requires the extra **n_iter** argument, which determines how many random cells must be selected. This determines how many times k-fold cross-validation will be performed. Therefore, by choosing a smaller number, fewer hyperparameter combinations will be considered and the method will take less time to complete. Also, please note that the **param_grid** argument is changed to **param_distributions** here. The **param_distributions** argument can take a dictionary with parameter names as keys, and either list of parameters or distributions as values for each key.

It could be argued that **RandomizedSearchCV()** is not as good as **GridSearchCV()** since it does not consider all the possible values and combinations of values for hyperparameters, which is reasonable. As a result, one smart way of performing hyperparameter tuning for deep learning models is to start with either **RandomizedSearchCV()** on many hyperparameters, or **GridSearchCV()** on fewer hyperparameters with larger gaps between them.

By beginning with a randomized search on many hyperparameters, we can determine which hyperparameters have the most influence on a model's performance. It can also help narrow down the range for important hyperparameters. Then, you can complete your hyperparameter tuning by performing `GridSearchCV()` on the smaller number of hyperparameters and the smaller ranges for each of them. This is called the **coarse-to-fine** approach to hyperparameter tuning.

Now, you are ready to practice implementing hyperparameter tuning using scikit-learn optimizers. In the next activity, you will try to improve your model for the diabetes dataset by tuning the hyperparameters.

ACTIVITY 5.03: HYPERPARAMETER TUNING ON THE AVILA PATTERN CLASSIFIER

The Avila dataset has been extracted from **800** images of the Avila Bible, a giant 12th-century Latin copy of the Bible. The dataset consists of various features about the images of the text, such as intercolumnar distance and margins of the text. The dataset also contains a class label that indicates if the pattern of the image falls into the most frequently occurring category or not. In this activity, you will build a Keras model similar to those in the previous activities, but this time, you will add regularization methods to your model as well. Then, you will use scikit-learn optimizers to perform tuning on the model hyperparameters, including the hyperparameters of the regularizers. Here are the steps you need to complete in this activity:

1. Load the dataset from the **data** subfolder of the **Chapter05** folder from GitHub using `X = pd.read_csv('../data/avila-tr_feats.csv')` and `y = pd.read_csv('../data/avila-tr_target.csv')`.

2. Define a function that returns a Keras model with three hidden layers, the first of `size 10`, the second of `size 6`, and the third of `size 4`, all with `L2 weight regularizations`. Use these values as the hyperparameters for your model: `activation='relu'`, `loss='binary_crossentropy'`, `optimizer='sgd'`, and `metrics=['accuracy']`. Also, make sure to pass the `L2 lambda` hyperparameter as an argument to your function so that we can tune it later.

3. Create the wrapper for your Keras model and perform **GridSearchCV()** on it using **cv=5**. Then, add the following values in the parameter grid: **lambda_parameter = [0.01, 0.5, 1]**, **epochs = [50, 100]**, and **batch_size = [20]**. This might take some time to process. Once the parameter search is complete, print the accuracy and the hyperparameters of the best cross-validation score. You can also print every other cross-validation score, along with the hyperparameters that resulted in that score.

4. Repeat the previous step, this time using **GridSearchCV()** on a narrower range with **lambda_parameter = [0.001, 0.01, 0.05, 0.1]**, **epochs = [400]**, and **batch_size = [10]**. It might take some time to process.

5. Repeat the previous step, but remove the **L2 regularizers** from your Keras model and instead of adding dropout regularization with the **rate** parameter at each hidden layer. Perform **GridSearchCV()** on the model using the following values in the parameter grid and print the results: **rate = [0, 0.2, 0.4]**, **epochs = [350, 400]**, and **batch_size = [10]**.

6. Repeat the previous step using **rate = [0.0, 0.05, 0.1]** and **epochs=[400]**.

After implementing these steps, you should see the following expected output:

```
Best cross-validation score= 0.7862895488739013
Parameters for Best cross-validation score= {'batch_size': 20, 'epochs':
100, 'rate': 0.0}
Accuracy 0.786290 (std 0.013557) for params {'batch_size': 20, 'epochs':
100, 'rate': 0.0}
Accuracy 0.786098 (std 0.005184) for params {'batch_size': 20, 'epochs':
100, 'rate': 0.05}
Accuracy 0.772004 (std 0.013733) for params {'batch_size': 20, 'epochs':
100, 'rate': 0.1}
```

NOTE

The solution for this activity can be found on page 422.

In this activity, we learned how to implement hyperparameter tuning on a Keras model with regularizers to perform classification using a real-life dataset. We learned how to use scikit-learn optimizers to perform tuning on model hyperparameters, including the hyperparameters of the regularizers. In this section, we implemented hyperparameter tuning by creating a grid of hyperparameters and iterating through them. This allows us to find the optimal set of hyperparameters using a scikit-learn workflow.

SUMMARY

In this chapter, you learned about two very important groups of techniques for improving the accuracy of your deep learning models: regularization and hyperparameter tuning. You learned how regularization helps address the overfitting problem by means of several different methods, including L1 and L2 norm regularization and dropout regularization—the more commonly used regularization techniques. You discovered the importance of hyperparameter tuning for machine learning models and the challenge of hyperparameter tuning for deep learning models in particular. You even practiced using scikit-learn optimizers to perform hyperparameter tuning on Keras models.

In the next chapter, you will explore the limitations of accuracy metrics when evaluating model performance, as well as other metrics (such as **precision**, **sensitivity**, **specificity**, and **AUC-ROC score**), including how to use them in order to gauge the quality of your model's performance better.

6

MODEL EVALUATION

OVERVIEW

This chapter covers model evaluation in depth. We will discuss alternatives to accuracy to evaluate the performance of a model when standard techniques are not feasible, especially where there are imbalanced classes. Finally, we will utilize confusion matrices, sensitivity, specificity, precision, FPR, ROC curves, and AUC scores to evaluate the performance of classifiers. By the end of this chapter, you will have an in-depth understanding of accuracy and null accuracy and will be able to understand and combat the challenges of imbalanced datasets.

INTRODUCTION

In the previous chapter, we covered **regularization** techniques for neural networks. **Regularization** is an important technique when it comes to combatting how a model overfits the training data and helps the model perform well on new, unseen data examples. One of the regularization techniques we covered involved **L1** and **L2** weight regularizations, in which penalization is added to the weights. The other regularization technique we learned about was **dropout regularization**, in which some units of layers are randomly removed from the model fitting process at each iteration. Both regularization techniques are designed to prevent individual weights or units by influencing them too strongly and allowing them to generalize as well.

In this chapter, we will learn about some different evaluation techniques other than **accuracy**. For any data scientist, the first step after building a model is to evaluate it, and the easiest way to evaluate a model is through its accuracy. However, in real-world scenarios, particularly where there are classification tasks with highly imbalanced classes such as for predicting the presence of hurricanes, predicting the presence of a rare disease, or predicting if someone will default on a loan, evaluating the model using its accuracy score is not the best evaluation technique.

This chapter explores core concepts such as imbalanced datasets and how different evaluation techniques can be used to work through these imbalanced datasets. This chapter begins with an introduction to accuracy and its limitations. Then, we will explore the concepts of **null accuracy**, **imbalanced datasets**, **sensitivity**, **specificity**, **precision**, **false positives**, ROC **curves**, and **AUC scores**.

ACCURACY

To understand accuracy properly, let's explore model evaluation. Model evaluation is an integral part of the model development process. Once you've built your model and executed it, the next step is to evaluate your model.

A model is built on a **training dataset** and evaluating a model's performance on the same training dataset is bad practice in data science. Once a model has been trained on a training dataset, it should be evaluated on a dataset that is completely different from the training dataset. This dataset is known as the **test dataset**. The objective should always be to build a model that generalizes, which means the model should produce similar (but not the same) results, or relatively similar results, on any dataset. This can only be achieved if we evaluate the model on data that is unknown to it.

The model evaluation process requires a metric that can quantify a model's performance. The simplest metric for model evaluation is accuracy. **Accuracy** is the fraction of predictions that our model gets right. This is the formula for calculating **accuracy**:

Accuracy = (Number of correct predictions) / (Total number of predictions)

For example, if we have **10** records and **7** are predicted correctly, then we can say that the accuracy of our model is **70%**. This is calculated as **7/10 = 0.7** or **70%**.

Null accuracy is the accuracy that can be achieved by predicting the most frequent class. If we don't run an algorithm and just predict accuracy based on the most frequent outcome, then the accuracy that's calculated based on this prediction is known as **null accuracy**:

Null accuracy = (Total number of instances of the frequently occurring class) / (Total number of instances)

Take a look at this example:

10 actual outcomes: [1,0,0,0,0,0,0,0,1,0].

Prediction: [0,0,0,0,0,0,0,0,0,0]

Null accuracy = 8/10 = 0.8 or 80%

So, our null accuracy is **80%**, meaning we are correct **80%** of the time. This means we have achieved **80%** accuracy without running an algorithm. Always remember that when null accuracy is high, it means that the distribution of response variables is skewed in favor of the frequently occurring class.

Let's work on an exercise to find the null accuracy of a dataset. The null accuracy of a dataset can be found by using the **value_count** function in the pandas library. The **value_count** function returns a series containing counts of unique values.

> **NOTE**
>
> All the Jupyter Notebooks for the exercises and activities in this chapter are available on GitHub at https://packt.live/37jHNUR.

EXERCISE 6.01: CALCULATING NULL ACCURACY ON A PACIFIC HURRICANES DATASET

We have a dataset documenting whether a **hurricane** has been observed in the Pacific Ocean that has two columns, **Date** and **hurricane**. The **Date** column indicates the date of the observation, while the **hurricane** column indicates whether there was a hurricane on that date. Rows with a **hurricane** value of **1** means there was a hurricane, while **0** means there was no hurricane. Find the **null accuracy** of the dataset by following these steps:

1. Open a Jupyter notebook. Import all the required libraries and load the **pacific_hurricanes.csv** file into the **data** folder from this book's GitHub repository:

```
# Import the data
import pandas as pd
df = pd.read_csv("../data/pacific_hurricanes.csv")
df.head()
```

The following is the output of the preceding code:

	Date	hurricane
0	1949-06-11	0
1	1949-06-12	0
2	1949-06-13	0
3	1949-06-14	0
4	1949-06-15	0

Figure 6.1: Data exploration of the pacific hurricanes dataset

2. Use the built-in **value_count** function from the pandas library to get the distribution for the data of the **hurricane** column. The **value_count** function shows the total instances of unique values:

```
df['hurricane'].value_counts()
```

The preceding code produces the following output:

```
0 22435
1 1842
Name: hurricane, dtype: int64
```

3. Use the **value_count** function and set the **normalize** parameter to **True**. To find the null accuracy, you will have to index the **pandas** series that was produced for index **0** to get the proportion of values related to no hurricanes occurring on a given day:

```
df['hurricane'].value_counts(normalize=True).loc[0]
```

The preceding code produces the following output:

```
0.9241257156979857
```

The calculated **null accuracy** of the dataset is **92.4126%**.

Here, we can see that our dataset has a very high null accuracy of **92.4126%**. So, if we just make a dumb model that predicts the majority class for all outcomes, our model will be **92.4126%** accurate.

> **NOTE**
>
> To access the source code for this specific section, please refer to https://packt.live/31FtQBm.
>
> You can also run this example online at https://packt.live/2ArNwNT.

Later in this chapter, in *Activity 6.01, Computing the Accuracy and Null Accuracy of a Neural Network When We Change the Train/Test Split*, we will see how null accuracy changes as we change the **test/train** split.

ADVANTAGES AND LIMITATIONS OF ACCURACY

The advantages of accuracy are as follows:

- **Easy to use**: Accuracy is very easy to compute and understand as it is just a simple fraction formula.

- **Popular compared to other techniques**: Since it is the easiest metric to compute, it is also the most popular and is universally accepted as the first step of evaluating a model. Most introductory books on data science teach accuracy as an evaluation metric.

- **Good for comparing different models**: Suppose you are trying to solve a problem with different models. You can always trust the model that gives the highest accuracy.

The limitations of accuracy are as follows:

- **No representation of response variable distribution**: Accuracy doesn't give us an idea of the distribution of the **response/dependent** variable. If we get an accuracy of **80%** in our model, we have no idea how the response variable is distributed and what the null accuracy of the dataset is. If the null accuracy of our dataset is above **70%**, then an **80%** accurate model is pretty useless.

- **Type 1 and type 2 errors**: `Accuracy` also gives us no information about `type 1` and `type 2` errors of the model. A `type 1` error is when a class is `negative` and we have predicted it as `positive`, while a `type 2` error is when a class is positive and we have predicted it as negative. We will be studying both of these errors later in this chapter. In the next section, we will cover the imbalanced datasets. The accuracy scores for models classifying imbalanced datasets can be particularly misleading, which is why other evaluation metrics are useful for model evaluation.

IMBALANCED DATASETS

Imbalanced datasets are a distinct case for classification problems where the class distribution varies between the classes. In such datasets, one class is overwhelmingly dominant. In other words, the `null accuracy` of an imbalanced dataset is very high.

Consider an example of credit card fraud. If we have a dataset of credit card transactions, then we will find that, of all the transactions, a very minuscule number of transactions were fraudulent and the majority of transactions were normal transactions. If **1** represents a fraudulent transaction and **0** represents a normal transaction, then there will be many 0s and hardly any 1s. The `null accuracy` of the dataset may be more than **99%**. This means that the majority class (in this case, **0**) is overwhelmingly greater than the minority class (in this case, **1**). Such sets are imbalanced datasets. Consider the following figure, which shows a general imbalanced dataset `scatter plot`:

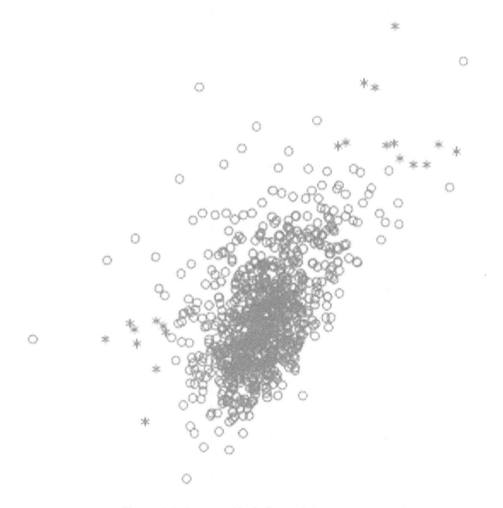

Figure 6.2: A general imbalanced dataset scatter plot

The preceding plot shows a generalized scatter plot of an imbalanced dataset, where the stars represent the minority class and the circles represent the majority class. As we can see, there are many more circles than stars; this can make it difficult for machine learning models to distinguish between the two classes. In the next section, we will cover some approaches to working with imbalanced datasets.

WORKING WITH IMBALANCED DATASETS

In machine learning, there are two ways of overcoming the shortcomings of imbalanced datasets, which are as follows:

- **Sampling techniques**: One way we can mitigate the imbalance of a dataset is by using special sampling techniques with which we can select our training and testing data in such a way that there is an adequate representation of all the classes. There are many such techniques—for instance, oversampling the minority class (meaning we take more samples from the minority class) or undersampling the majority class (meaning we take a smaller sample from the majority class). However, if the data is highly imbalanced with null accuracies above **90%**, then sampling techniques struggle to give the correct representation of majority-minority classes in the data and our model may overfit. So, the best way is to modify our evaluation techniques.

- **Modifying model evaluation techniques**: When working with highly imbalanced datasets, it is better to modify model evaluation techniques. This is the most robust way to get good results, which means using these methods will likely achieve good results on new, unseen data. There are many evaluation metrics other than accuracy that can be modified to evaluate a model. To learn about all those techniques, it is important to understand the concept of the confusion matrix.

CONFUSION MATRIX

A **confusion matrix** describes the performance of the classification model. In other words, a confusion matrix is a way to summarize classifier performance. The following table shows a basic representation of a confusion matrix and represents how the predicted results by the model compared to the true values:

	Predicted 0	Predicted 1
Actual 0	TN	FP
Actual 1	FN	TP

Figure 6.3: Basic representation of a confusion matrix

Let's go over the meanings of the abbreviations that were used in the preceding table:

- **TN** (**True negative**): This is the count of outcomes that were originally negative and were predicted negative.

- **FP** (**False positive**): This is the count of outcomes that were originally negative but were predicted positive. This error is also called a **type 1 error**.

- **FN** (**False negative**): This is the count of outcomes that were originally positive but were predicted negative. This error is also called a **type 2 error**.

- **TP** (**True positive**): This is the count of outcomes that were originally positive and were predicted as positive.

The goal is to maximize the values in the **TN** and **TP** boxes in the preceding table, that is, the true negatives and true positives, and minimize the values in the **FN** and **FP** boxes, that is, the false negatives and false positives.

The following code is an example of a confusion matrix:

```
from sklearn.metrics import confusion_matrix
cm = confusion_matrix(y_test,y_pred_class)
print(cm)
```

The preceding code produces the following output:

```
array([[89, 2],
       [13, 4]], dtype=int64)
```

The aim of all machine learning and deep learning algorithms is to maximize TN and TP and minimize FN and FP. The following example code calculates TN, FP, FN, and TP:

```
# True Negative
TN = cm[0,0]
# False Negative
FN = cm[1,0]
# False Positives
FP = cm[0,1]
# True Positives
TP = cm[1,1]
```

> **NOTE**
>
> Accuracy does not help us understand type 1 and type 2 errors.

METRICS COMPUTED FROM A CONFUSION MATRIX

The metrics that can be derived from a **confusion matrix** are **sensitivity**, **specificity**, **precision**, **FP rate**, ROC, and **AUC**:

- **Sensitivity**: This is the number of positive predictions divided by the total actual number of positives. Sensitivity is also known as recall or true positive. In our case, it is the total number of patients classified as **1**, divided by the total number of patients who are actually **1**:

  Sensitivity = TP / (TP+FN)

Sensitivity refers to how often the prediction is correct when the actual value is positive. In cases such as building a model to predict patient readmission at a hospital, we need our model to be highly sensitive. We need 1 to be predicted as **1**. If a **0** is predicted as **1**, it is acceptable, but if a **1** is predicted as **0**, it means a patient who was readmitted is predicted as not readmitted, and this will cause severe penalties for the hospital.

- **Specificity**: This is the number of negative predictions divided by the total number of actual negatives. To use the previous example, it would be readmission predicted as **0** divided by the total number of patients who were actually **0**. Specificity is also known as the true negative rate:

Specificity = TN / (TN+FP)

Specificity refers to how often the prediction is correct when the actual value is negative. There are cases, such as spam email detection, where we need our algorithm to be more specific. The model predicts **1** when an email is spam and **0** when it isn't. We want the model to predict **0** as always **0**, because if a non-spam email is classified as spam, important emails may end up in the spam folder. Sensitivity can be compromised here because some spam emails may arrive in our inbox, but non-spam emails should never go to the spam folder.

> **NOTE**
>
> As we discussed previously, whether a model should be sensitive or specific totally depends on the business problem.

- **Precision**: This is the true positive prediction divided by the total number of positive predictions. Precision refers to how often we are correct when the value predicted is positive:

Precision = TP / (TP+FP)

- **False Positive Rate** (**FPR**): The **FPR** is calculated as the ratio between the number of false-positive events and the total number of actual negative events. **FPR** refers to how often we are incorrect when the actual value is negative. **FPR** is also equal to **1** - specificity:

False positive rate = FP / (FP+TN)

- **Receiver Operating Characteristic (ROC) curve**: Another important way to evaluate a classification model is by using a **ROC curve**. A **ROC curve** is a plot between the true positive rate (**sensitivity**) and the **FPR(1 - specificity)**. The following plot shows an example of an **ROC curve**:

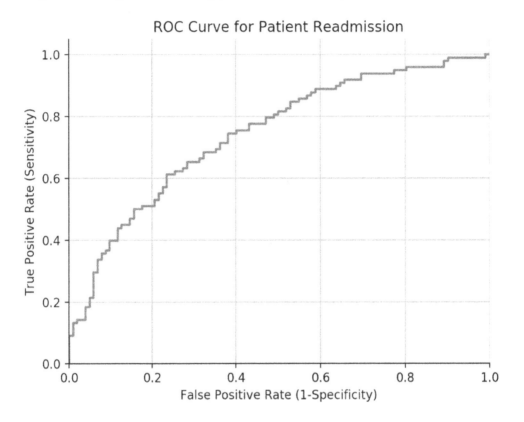

Figure 6.4: An example of ROC curve

To decide which **ROC curve** is the best among multiple curves, we need to look at the empty space on the upper left of the curve—the smaller the space, the better the result. The following plot shows an example of multiple **ROC curves**:

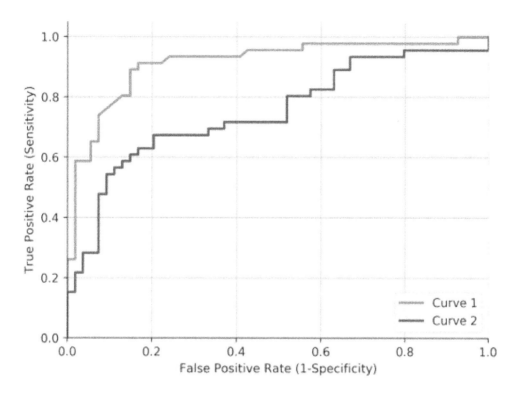

Figure 6.5: An example of multiple ROC curves

NOTE

The red curve is better than the blue curve because it leaves less space in the upper-left corner.

The **ROC curve** of a model tells us the relationship between **sensitivity** and **specificity**.

- **Area Under Curve** (**AUC**): This is the area under the **ROC curve**. Sometimes, **AUC** is also written as **AUROC**, meaning the area under the **ROC curve**. Basically, **AUC** is a numeric value that represents the area under a **ROC curve**. The larger the area under the **ROC**, the better, and the bigger the **AUC score**, the better. The preceding plot shows us an example of an **AUC**.

In the preceding plot, the **AUC** of the red curve is greater than the **AUC** of the blue curve, which means the **AUC** of the red curve is better than the AUC of the blue curve. There is no standard rule for the **AUC score**, but here are some generally acceptable values and how they relate to model quality:

AUC Score	Model Quality
0.9 to 1	Excellent
0.8 to 0.9	Good
0.7 to 0.8	Fair
0.6 to 0.7	Poor
0.5 to 0.6	Fail

Figure 6.6: General acceptable AUC score

Now that we understand the theory behind the various metrics, let's complete some activities and exercises to implement what we have learned.

EXERCISE 6.02: COMPUTING ACCURACY AND NULL ACCURACY WITH APS FAILURE FOR SCANIA TRUCKS DATA

The dataset that we will be using in this exercise consists of data that's been collected from heavy Scania trucks in everyday usage that have failed in some way. The system in focus is the **Air pressure system** (APS), which generates pressurized air that is utilized in various functions in a truck, such as braking and gear changes. The positive class in the dataset represents component failures for a specific component in the APS, while the negative class represents failures for components not related to the APS.

The objective of this exercise is to predict which trucks have had failures due to the APS so that the repair and maintenance mechanics have the information they can work with when checking why the truck failed and which area of the truck needs to be inspected.

> **NOTE**
>
> The dataset for this exercise can be downloaded from this book's GitHub repository at https://packt.live/2SGEEsH.
>
> Throughout this exercise, you may get slightly different results due to the random nature of the internal mathematical operations.

Data preprocessing and exploratory data analysis:

1. Import the required libraries. Load the dataset using the pandas **read_csv** function and explore the first **five** rows of the dataset:

```
#import the libraries
import numpy as np
import pandas as pd


# Load the Data
X = pd.read_csv("../data/aps_failure_training_feats.csv")
y = pd.read_csv("../data/aps_failure_training_target.csv")


# use the head function view the first 5 rows of the data
X.head()
```

The following table shows the output of the preceding code:

	aa_000	ab_000	ac_000	ad_000	ae_000	af_000	ag_000	ag_001	ag_002	ag_003	...	ee_002	ee_003	ee_004	ee_005	ee_006	ee_007
0	76698	0.0	2.130706e+09	280.0	0.0	0.0	0.0	0.0	0.0	0.0	...	1240520.0	493384.0	721044.0	469792.0	339156.0	157956.0
1	33058	0.0	0.000000e+00	0.0	0.0	0.0	0.0	0.0	0.0	0.0	...	421400.0	178064.0	293306.0	245416.0	133654.0	81140.0
2	41040	0.0	2.280000e+02	100.0	0.0	0.0	0.0	0.0	0.0	0.0	...	277378.0	159812.0	423992.0	409564.0	320746.0	158022.0
3	12	0.0	7.000000e+01	66.0	0.0	10.0	0.0	0.0	0.0	318.0	...	240.0	46.0	58.0	44.0	10.0	0.0
4	60874	0.0	1.368000e+03	458.0	0.0	0.0	0.0	0.0	0.0	0.0	...	622012.0	229790.0	405298.0	347188.0	286954.0	311560.0

5 rows × 170 columns

Figure 6.7: First five rows of the patient readmission dataset

2. Describe the feature values in the dataset using the **describe** method:

```
# Summary of Numerical Data
X.describe()
```

The following table shows the output of the preceding code:

	aa_000	ab_000	ac_000	ad_000	ae_000	af_000	ag_000	ag_001	ag_002	ag_003	...
count	6.000000e+04	60000.000000	6.000000e+04	6.000000e+04	60000.000000	60000.000000	6.000000e+04	6.000000e+04	6.000000e+04	6.000000e+04	...
mean	5.933650e+04	0.162500	3.362258e+08	1.434071e+05	6.535000	10.548200	2.191577e+02	9.648104e+02	8.509771e+03	8.760054e+04	...
std	1.454301e+05	1.687318	7.767625e+08	3.504525e+07	158.147893	205.387115	2.036364e+04	3.400891e+04	1.494818e+05	7.575171e+05	...
min	0.000000e+00	0.000000	0.000000e+00	0.000000e+00	0.000000	0.000000	0.000000e+00	0.000000e+00	0.000000e+00	0.000000e+00	...
25%	8.340000e+02	0.000000	8.000000e+00	0.000000e+00	0.000000	0.000000	0.000000e+00	0.000000e+00	0.000000e+00	0.000000e+00	...
50%	3.077600e+04	0.000000	1.200000e+02	4.200000e+01	0.000000	0.000000	0.000000e+00	0.000000e+00	0.000000e+00	0.000000e+00	...
75%	4.866800e+04	0.000000	8.480000e+02	2.920000e+02	0.000000	0.000000	0.000000e+00	0.000000e+00	0.000000e+00	0.000000e+00	...
max	2.746564e+06	204.000000	2.130707e+09	8.584298e+09	21050.000000	20070.000000	3.376892e+06	4.109372e+06	1.055286e+07	6.340207e+07	...

8 rows × 170 columns

Figure 6.8: Numerical metadata of the patient readmission dataset

> **NOTE**
>
> Independent variables are also known as explanatory variables, while dependent variables are also known as **response variables**. Also, remember that indexing in Python starts from **0**.

3. Explore **y** using the **head** function:

```
y.head()
```

The following table shows the output of the preceding code:

	class
0	0
1	0
2	0
3	0
4	0

Figure 6.9: The first five rows of the y variable of the patient readmission dataset

4. Split the data into test and train sets by using the **train_test_split** function from the scikit-learn library. To make sure we all get the same results, set the **random_state** parameter to **42**. The data is split with an **80:20 ratio**, meaning **80%** of the data is **training data** and the remaining **20%** is **test data**:

```
from sklearn.model_selection import train_test_split
seed = 42
X_train, X_test, \
y_train, y_test= train_test_split(X, y, test_size=0.20, \
                                  random_state=seed)
```

5. Scale the training data using the **StandardScaler** function and use the scaler to scale the **test data**:

```
# Initialize StandardScaler
from sklearn.preprocessing import StandardScaler
sc = StandardScaler()

# Transform the training data
X_train = sc.fit_transform(X_train)
X_train = pd.DataFrame(X_train,columns=X_test.columns)

# Transform the testing data
X_test = sc.transform(X_test)
X_test = pd.DataFrame(X_test,columns=X_train.columns)
```

> **NOTE**
>
> The **sc.fit_transform()** function transforms the data and the data is converted into a **NumPy** array. We may need the data for further analysis in the DataFrame objects, so the **pd.DataFrame()** function reconverts data into a DataFrame.

This completes the data preprocessing part of this exercise. Now, we need to build a neural network and calculate the **accuracy**.

6. Import the libraries that are required for creating the neural network architecture:

```
# Import the relevant Keras libraries
from keras.models import Sequential
from keras.layers import Dense
from keras.layers import Dropout
from tensorflow import random
```

7. Initiate the **Sequential** class:

```
# Initiate the Model with Sequential Class
np.random.seed(seed)
random.set_seed(seed)
model = Sequential()
```

8. Add **five** hidden layers of the **Dense** class and the add **Dropout** after each. Build the first hidden layer so that it has a size of **64** and with a dropout rate of **0.5**. The second hidden layer will have a size of **32** and a dropout rate of **0.4**. The third hidden layer will have a size of **16** and a dropout rate of **0.3**. The fourth hidden layer will have a size of **8** and dropout rate of **0.2**. The final hidden layer will have a size of **4** and a dropout rate of **0.1**. Each hidden layer will have a **ReLU activation** function and the kernel initializer will be set to **uniform**:

```
# Add the hidden dense layers and with dropout Layer
model.add(Dense(units=64, activation='relu', \
                kernel_initializer='uniform', \
                input_dim=X_train.shape[1]))
model.add(Dropout(rate=0.5))
model.add(Dense(units=32, activation='relu', \
                kernel_initializer='uniform'))
model.add(Dropout(rate=0.4))
model.add(Dense(units=16, activation='relu', \
                kernel_initializer='uniform'))
model.add(Dropout(rate=0.3))
model.add(Dense(units=8, activation='relu', \
                kernel_initializer='uniform'))
model.add(Dropout(rate=0.2))
model.add(Dense(units=4, activation='relu', \
                kernel_initializer='uniform'))
model.add(Dropout(rate=0.1))
```

9. Add an output **Dense** layer with a **sigmoid** activation function:

```
# Add Output Dense Layer
model.add(Dense(units=1, activation='sigmoid', \
                kernel_initializer='uniform'))
```

> **NOTE**
>
> Since the output is binary, we are using the **sigmoid** function. If the output
> is multiclass (that is, more than two classes), then the **softmax** function
> should be used.

10. Compile the network and fit the model. Calculate the accuracy during the training process by setting **metrics=['accuracy']**:

```
# Compile the model
model.compile(optimizer='adam', \
              loss='binary_crossentropy', \
              metrics=['accuracy'])
```

11. Fit the model with **100** epochs, a batch size of **20**, and a validation split of **20%**:

```
#Fit the Model
model.fit(X_train, y_train, epochs=100, \
          batch_size=20, verbose=1, \
          validation_split=0.2, shuffle=False)
```

12. Evaluate the model on the **test** dataset:

```
test_loss, test_acc = model.evaluate(X_test, y_test)
print(f'The loss on the test set is {test_loss:.4f} \
and the accuracy is {test_acc*100:.4f}%')
```

The preceding code produces the following output:

```
12000/12000 [==============================] - 0s 20us/step
The loss on the test set is 0.0802 and the accuracy is 98.9917%
```

The model returns an accuracy of **98.9917%**. But is it good enough? We can only get the answer to this question by comparing it with the null accuracy.

Compute the null accuracy:

13. The null accuracy can be calculated using the **value_count** function of the pandas library, which we used in *Exercise 6.01, Calculating Null Accuracy on a Pacific Hurricanes Dataset*, of this chapter:

```
"""
Use the value_count function to calculate distinct class values
"""
y_test['class'].value_counts()
```

The preceding code produces the following output:

```
0    11788
1      212
Name: class, dtype: int64
```

14. Calculate the **null** accuracy:

```
# Calculate the null accuracy
y_test['class'].value_counts(normalize=True).loc[0]
```

The preceding code produces the following output:

```
0.9823333333333333
```

Here, we have obtained the null accuracy of the model. As we conclude this exercise, the following points must be noted: the accuracy of our model is **98.9917%**, approximately. Under ideal conditions, **98.9917%** accuracy is very **good** accuracy, but here, the **null accuracy** is **very high**, which helps put our model's performance into perspective. The **null accuracy** of our model is **98.2333%**. Since the **null accuracy** of the model is **so high**, an **accuracy** of **98.9917%** is not significant but certainly respectable, and **accuracy** in such cases is not the correct metric to evaluate an algorithm with.

> **NOTE**
>
> To access the source code for this specific section, please refer to https://packt.live/31FUb2d.
>
> You can also run this example online at https://packt.live/3goL0ax.

Now, let's go through activity on computing the accuracy and null accuracy of the neural network model when we change the train/test split.

ACTIVITY 6.01: COMPUTING THE ACCURACY AND NULL ACCURACY OF A NEURAL NETWORK WHEN WE CHANGE THE TRAIN/TEST SPLIT

A train/test split is a random sampling technique. In this activity, we will see that our null accuracy and accuracy will be affected by changing the **train/test** split. To implement this, the part of the code where the train/test split was defined has to be changed. We will use the same dataset that we used in *Exercise 6.02, Computing Accuracy and Null Accuracy with APS Failure for Scania Trucks Data*. Follow these steps to complete this activity:

1. Import all the necessary dependencies and load the dataset.

2. Change **test_size** and **random_state** from 0.20 to 0.30 and 42 to 13, respectively.

3. Scale the data using the **StandardScaler** function.

4. Import the libraries that are required to build a neural network architecture and initiate the **Sequential** class.

5. Add the **Dense** layers with **Dropout**. Set the first hidden layer so that it has a size of **64** with a dropout rate of 0.5, the second hidden layer so that it has a size of **32** with a dropout rate of 0.4, the third hidden layer so that is has a size of **16** with a dropout rate of 0.3, the fourth hidden layer so that it has a size of **8** with a dropout rate of 0.2, and the final hidden layer so that it has a size of **4** with a dropout rate of 0.1. Set all the activation functions to **ReLU**.

6. Add an output **Dense** layer with the **sigmoid** activation function.

7. Compile the network and fit the model using accuracy. Fit the model with 100 epochs and a batch size of 20.

8. Fit the model to the training data while saving the results from the fit process.

9. Evaluate the model on the test dataset.

10. Count the number of values in each class of the test target dataset.

11. Calculate the null accuracy using the pandas **value_count** function.

> **NOTE**
>
> In this activity, you may get slightly different results due to the random nature of internal mathematical operations.

Here, we can see that the accuracy and null accuracy will change as we change the **train/test** split. We will not cover any sampling techniques in this chapter as we have a very highly imbalanced dataset, and sampling techniques will not yield any fruitful results.

> **NOTE**
>
> The solution for this activity can be found on page 430.

Let's move on to the next exercise and compute the metrics that have been derived from the confusion matrix.

EXERCISE 6.03: DERIVING AND COMPUTING METRICS BASED ON A CONFUSION MATRIX

The dataset that we will be using in this exercise consists of data that has been collected from heavy Scania trucks in everyday usage that have failed in some way. The system that's in focus is the **Air Pressure System** (APS), which generates pressurized air that is utilized in various functions in a truck, such as braking and gear changes. The positive class in the dataset represents component failures for a specific component in the APS, while the negative class represents failures for components not related to the APS.

The objective of this exercise is to predict which trucks have had failures due to the APS, much like we did in the previous exercise. We will derive the sensitivity, specificity, precision, and false positive rate of the neural network model to evaluate its performance. Finally, we will adjust the threshold value and recompute the sensitivity and specificity. Follow these steps to complete this exercise:

> **NOTE**
>
> The dataset for this exercise can be downloaded from this book's GitHub repository at https://packt.live/2SGEEsH.
>
> You may get slightly different results due to the random nature of internal mathematical operations.

1. Import the necessary libraries and load the data using the pandas **read_csv** function:

```
# Import the libraries
import numpy as np
import pandas as pd


# Load the Data
X = pd.read_csv("../data/aps_failure_training_feats.csv")
y = pd.read_csv("../data/aps_failure_training_target.csv")
```

2. Next, split the data into training and test datasets using the **train_test_split** function:

```
from sklearn.model_selection import train_test_split
seed = 42
X_train, X_test, \
y_train, y_test = train_test_split(X, y, \
                    test_size=0.20, random_state=seed)
```

3. Following this, scale the feature data so that it has a **mean** of **0** and a **standard deviation** of **1** using the **StandardScaler** function. Fit the scaler to the **training data** and apply it to the **test data**:

```
from sklearn.preprocessing import StandardScaler
sc = StandardScaler()

# Transform the training data
X_train = sc.fit_transform(X_train)
X_train = pd.DataFrame(X_train,columns=X_test.columns)

# Transform the testing data
X_test = sc.transform(X_test)
X_test = pd.DataFrame(X_test,columns=X_train.columns)
```

4. Next, import the **Keras** libraries that are required to create the model. Instantiate a **Keras** model of the **Sequential** class and add five hidden layers to the model, including dropout for each layer. The first hidden layer should have a size of **64** and a dropout rate of **0.5**. The second hidden layer should have a size of **32** and a dropout rate of **0.4**. The third hidden layer should have a size of **16** and a dropout rate of **0.3**. The fourth hidden layer should have a size of **8** and a dropout rate of **0.2**. The final hidden layer should have a size of **4** and a dropout rate of **0.1**. All the hidden layers should have **ReLU activation** functions and have **kernel_initializer = 'uniform'**. Add a final output layer to the model with a **sigmoid activation** function. Compile the model by calculating the accuracy metric during the training process:

```
# Import the relevant Keras libraries
from keras.models import Sequential
from keras.layers import Dense
from keras.layers import Dropout
from tensorflow import random

np.random.seed(seed)
random.set_seed(seed)
model = Sequential()

# Add the hidden dense layers and with dropout Layer
```

```
model.add(Dense(units=64, activation='relu', \
                kernel_initializer='uniform', \
                input_dim=X_train.shape[1]))
model.add(Dropout(rate=0.5))
model.add(Dense(units=32, activation='relu', \
                kernel_initializer='uniform'))
model.add(Dropout(rate=0.4))
model.add(Dense(units=16, activation='relu', \
                kernel_initializer='uniform'))
model.add(Dropout(rate=0.3))
model.add(Dense(units=8, activation='relu', \
                kernel_initializer='uniform'))
model.add(Dropout(rate=0.2))
model.add(Dense(units=4, activation='relu', \
                kernel_initializer='uniform'))
model.add(Dropout(rate=0.1))

# Add Output Dense Layer
model.add(Dense(units=1, activation='sigmoid', \
                kernel_initializer='uniform'))

# Compile the Model
model.compile(optimizer='adam', \
              loss='binary_crossentropy', \
              metrics=['accuracy'])
```

5. Next, fit the model to the training data by training for **100** epochs with **batch_size=20** and **validation_split=0.2**:

```
model.fit(X_train, y_train, epochs=100, \
          batch_size=20, verbose=1, \
          validation_split=0.2, shuffle=False)
```

6. Once the model has finished fitting to the **training data**, create a variable that is the result of the model's prediction on the **test data** using the model's **predict** and **predict_proba** methods:

```
y_pred = model.predict(X_test)
y_pred_prob = model.predict_proba(X_test)
```

7. Next, compute the predicted class by setting the value of the prediction on the **test set** to **1** if the value is above **0.5** and **0** if it's below **0.5**. Compute the **confusion matrix** using the **confusion_matrix** function from **scikit-learn**:

```
from sklearn.metrics import confusion_matrix
y_pred_class1 = y_pred > 0.5
cm = confusion_matrix(y_test, y_pred_class1)
print(cm)
```

The preceding code produces the following output:

```
[[11730   58]
 [   69 143]]
```

Always use **y_test** as the first parameter and **y_pred_class1** as the second parameter so that you always get the correct results.

8. Calculate the true negative (**TN**), false negative (**FN**), false positive (**FP**), and true positive (**TP**):

```
# True Negative
TN = cm[0,0]

# False Negative
FN = cm[1,0]

# False Positives
FP = cm[0,1]

# True Positives
TP = cm[1,1]
```

> **NOTE**
>
> Using **y_test** and **y_pred_class1** in that order is necessary because if they are used in reverse order, the matrix will still be computed without errors, but will be incorrect.

9. Calculate the **sensitivity**:

```
# Calculating Sensitivity
Sensitivity = TP / (TP + FN)
print(f'Sensitivity: {Sensitivity:.4f}')
```

The preceding code produces the following output:

```
Sensitivity: 0.6745
```

10. Calculate the **specificity**:

```
# Calculating Specificity
Specificity = TN / (TN + FP)
print(f'Specificity: {Specificity:.4f}')
```

The preceding code produces the following output:

```
Specificity: 0.9951
```

11. Calculate the **precision**:

```
# Precision
Precision = TP / (TP + FP)
print(f'Precision: {Precision:.4f}')
```

The preceding code produces the following output:

```
Precision: 0.7114
```

12. Calculate the **false positive rate**:

```
# Calculate False positive rate
False_Positive_rate = FP / (FP + TN)
print(f'False positive rate: \
    {False_Positive_rate:.4f}')
```

The preceding code produces the following output:

```
False positive rate: 0.0049
```

The following image shows the output of the values:

Metric	Value
Sensitivity	0.6745 or 67.45%
Specificity	0.9951 or 99.51%
Precision	0.7114 or 71.14%
False Positive Rate	0.0049 or 0.049%

Figure 6.10: Metrics summary

NOTE

Sensitivity is inversely proportional to specificity.

As we discussed previously, our model should be more sensitive, but it looks more specific and less sensitive. So, how do we solve this? The answer lies in the threshold probabilities. The sensitivity of the model can be increased by adjusting the threshold value for classifying the dependent variable as **1** or **0**. Recall that, originally, we set the value of **y_pred_class1** to greater than **0.5**. Let's change the threshold to **0.3** and rerun the code to check the results.

13. Go to *step 7*, change the threshold from **0.5** to **0.3**, and rerun the code:

```
y_pred_class2 = y_pred > 0.3
```

14. Now, create a **confusion matrix** and calculate the **specificity** and **sensitivity**:

```
from sklearn.metrics import confusion_matrix
cm = confusion_matrix(y_test,y_pred_class2)
print(cm)
```

The preceding code produces the following output:

```
[[11700   88]
 [   58  154]]
```

For comparison, the following is the previous **confusion matrix** with a **threshold** of 0.5:

```
[[11730   58]
 [   69 143]]
```

> **NOTE**
>
> Always remember that the original values of **y_test** should be passed as the first parameter and **y_pred** as the second parameter.

15. Compute the various components of the **confusion matrix**:

```
# True Negative
TN = cm[0,0]

# False Negative
FN = cm[1,0]

# False Positives
FP = cm[0,1]

# True Positives
TP = cm[1,1]
```

16. Calculate the new **sensitivity**:

```
# Calculating Sensitivity
Sensitivity = TP / (TP + FN)
print(f'Sensitivity: {Sensitivity:.4f}')
```

The preceding code produces the following output:

```
Sensitivity: 0.7264
```

17. Calculate the **specificity**:

```
# Calculating Specificity
Specificity = TN / (TN + FP)
print(f'Specificity: {Specificity:.4f}')
```

The preceding code produces the following output:

```
Specificity: 0.9925
```

There is a clear increase in **sensitivity** and **specificity** after decreasing the threshold:

Threshold	Sensitivity	Specificity
0.5	67.45%	99.51%
0.3	72.64%	99.25%

Figure 6.11: Sensitivity and specificity comparison

So, clearly, decreasing the threshold value increases the sensitivity.

18. Visualize the data distribution. To understand why decreasing the threshold value increases the sensitivity, we need to see a histogram of our predicted probabilities. Recall that we created the **y_pred_prob** variable to predict the probabilities of the classifier:

```
import matplotlib.pyplot as plt
%matplotlib inline
# histogram of class distribution
plt.hist(y_pred_prob, bins=100)
plt.title("Histogram of Predicted Probabilities")
plt.xlabel("Predicted Probabilities of APS failure")
plt.ylabel("Frequency")
plt.show()
```

The following plot shows the output of the preceding code:

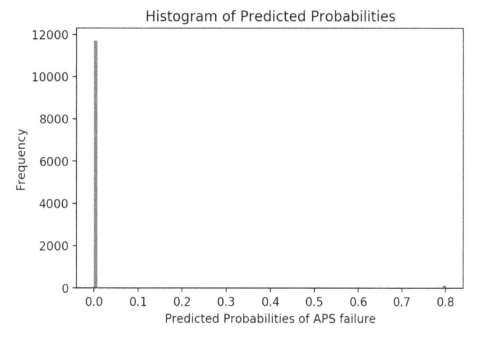

Figure 6.12: A histogram of the probabilities of patient readmission from the dataset

This histogram clearly shows that most of the probabilities for the predicted classifier lie in a range from **0.0** to **0.1**, which is indeed very low. Unless we set the threshold very low, we cannot increase the sensitivity of the model. Also, note that sensitivity is inversely proportional to specificity, so when one increases, the other decreases.

> **NOTE**
>
> To access the source code for this specific section, please refer to https://packt.live/31E6v32.
>
> You can also run this example online at https://packt.live/3gquh6y.

There is no universal value of the threshold, though the value of **0.5** is commonly used as a default. One method for selecting the threshold is to plot a histogram and then select the threshold manually. In our case, any threshold between **0.1** and **0.7** can be used as the model as there are few predictions between those values, as can be seen from the histogram that was produced at the end of the previous exercise.

Another method for choosing the threshold is to plot the **ROC curve**, which plots the true positive rate as a function of the false positive rate. Depending on your tolerance for each, the threshold value can be selected. Plotting the **ROC curve** is also a good technique if we wish to evaluate the performance of the model because the area under the **ROC curve** is a direct measure of the model's performance. In the next activity, we will explore the performance of our model using the **ROC curve** and the **AUC score**.

ACTIVITY 6.02: CALCULATING THE ROC CURVE AND AUC SCORE

The **ROC curve** and **AUC score** is an effective way to easily evaluate the performance of a binary classifier. In this activity, we will plot the **ROC curve** and calculate the **AUC score** of a model. We will use the same dataset and train the same model that we used in *Exercise 6.03*, *Deriving and Computing Metrics Based on a Confusion Matrix*. Use the APS failure data and calculate the **ROC curve** and **AUC score**. Follow these steps to complete this activity:

1. Import all the necessary dependencies and load the dataset.

2. Split the data into training and test datasets using the **train_test_split** function.

3. Scale the training and test data using the **StandardScaler** function.

4. Import the libraries that are required to build a neural network architecture and initiate the **Sequential** class. Add five **Dense** layers with **Dropout**. Set the first hidden layer so that it has a size of **64** with a dropout rate of **0.5**, the second hidden layer so that it has a size of **32** with a dropout rate of **0.4**, the third hidden layer so that it has a size of **16** with a dropout rate of **0.3**, the fourth hidden layer so that it has a size of **8** with a dropout rate of **0.2**, and the final hidden layer so that it has a size of **4**, with a dropout rate of **0.1**. Set all the activation functions to **ReLU**.

5. Add an output **Dense** layer with the **sigmoid** activation function. Compile the network then fit the model using accuracy. Fit the model with **100** epochs and a batch size of **20**.

6. Fit the model to the training data, saving the results from the fit process.

7. Create a variable representing the predicted classes of the test dataset.

8. Calculate the false positive rate and true positive rate using the **roc_curve** function from **sklearn.metrics**. The false positive rate and true positive rate are the first and second of three return variables. Pass the true values and the predicted values to the function.

9. Plot the ROC curve, which is the true positive rate as a function of the false positive rate.

10. Calculate the AUC score using the **roc_auc_score** from **sklearn.metrics** while passing the true values and predicted values of the model.

After implementing these steps, you should get the following output:

```
0.944787151628455
```

> **NOTE**
>
> The solution for this activity can be found on page 434.

In this activity, we learned how to calculate a **ROC** and an **AUC score** with the APS failure dataset. We also learned how specificity and sensitivity change with different threshold values.

SUMMARY

In this chapter, we covered model evaluation and accuracy in depth. We learned how accuracy is not the most appropriate technique for evaluation when our dataset is imbalanced. We also learned how to compute a confusion matrix using scikit-learn and how to derive other metrics, such as sensitivity, specificity, precision, and false positive rate.

Finally, we understood how to use threshold values to adjust metrics and how **ROC curves** and **AUC scores** help us evaluate our models. It is very common to deal with imbalanced datasets in real-life problems. Problems such as credit card fraud detection, disease prediction, and spam email detection all have imbalanced data in different proportions.

In the next chapter, we will learn about a different kind of neural network architecture (convolutional neural networks) that performs well on image classification tasks. We will test performance by classifying images into two classes and experiment with different architectures and activation functions.

7

COMPUTER VISION WITH CONVOLUTIONAL NEURAL NETWORKS

OVERVIEW

This chapter covers computer vision and how this is accomplished with neural networks. You will learn to build image processing applications and classify models with convolutional neural networks. You will also study the architecture of convolutional neural networks and how to utilize techniques such as max pooling and flattening, feature mapping, and feature detection. By the end of this chapter, you will be able to not only build your own image classifiers but also evaluate them effectively for your own applications.

INTRODUCTION

In the previous chapter, we explored model evaluation in detail. We covered **accuracy** and why it may be misleading for some datasets, especially for classification tasks with highly imbalanced classes. Datasets with imbalanced classes such as the prediction of hurricanes in the Pacific Ocean or the prediction of whether someone will default on their credit card loan have positive instances that are relatively rare compared to negative instances, so accuracy scores are misleading since the null accuracy is so high.

To combat class imbalance, we learned about techniques that we can use to appropriately evaluate our model, including calculating model evaluation metrics such as the sensitivity, specificity, false positive rate, and **AUC score**, and plotting the **ROC curve**. In this chapter, we will learn how to classify another type of dataset—namely, images. Image classification is extremely useful and there are many real-world applications of it, as we will discover.

Computer vision is one of the most important concepts in machine learning and artificial intelligence. With the wide use of smartphones for capturing, sharing, and uploading images every day, the amount of data that's generated through images is increasing exponentially. So, the need for experts who are specialized in the field of computer vision is at an all-time high. Industries such as the health care industry are on the verge of a revolution due to the progress that's been made in the field of medical imaging.

This chapter will introduce you to computer vision and the various industries in which computer vision is used. You will also learn about **Convolutional Neural Networks (CNNs)**, which are the most widely used neural networks for image processing. Like neural networks, CNNs are also made up of neurons that receive inputs that are processed using weighted sums and activation functions. However, unlike **ANNs**, which use vectors as inputs, CNN uses images as its input. In this chapter, we will be studying **CNNs** in greater detail, along with the associated concepts of **max pooling**, **flattening**, **feature maps**, and **feature selection**. We will use Keras as a tool to run image processing algorithms on real-life images.

COMPUTER VISION

To understand computer vision, let's discuss human vision. Human vision is the ability of the human eye and brain to see and recognize objects. Computer vision is the process of giving a machine a similar, if not better, understanding of seeing and identifying objects in the real world.

It is fairly simple for the human eye to precisely identify whether an animal is a tiger or a lion, but it takes a lot of training for a computer system to understand such objects distinctly. Computer vision can also be defined as building mathematical models that can mimic the function of a human eye and brain. Basically, it is about training computers to understand and process images and videos.

Computer vision is an integral part of many cutting-edge areas of robotics: health care and medical (X-rays, MRI scans, CT scans, and so on), drones, self-driving cars, sports and recreation, and so on. Almost all businesses need computer vision to run successfully.

Imagine a large amount of data that's generated by CCTV footage across the world, the number of pictures our smartphones capture each day, the number of videos that are shared on internet sites such as YouTube on a daily basis, and the pictures we share on popular social networking sites such as Facebook and Instagram. All of this generates huge volumes of image data. To process and analyze this data and make computers more intelligent in terms of processing, this data requires high-level experts who specialize in computer vision. Computer vision is a highly lucrative field in machine learning. The following sections will describe how computer vision is achieved with neural networks—and particularly convolutional neural networks—that perform well for computer vision tasks.

CONVOLUTIONAL NEURAL NETWORKS

When we talk about computer vision, we talk about CNNs in the same breath. CNN is a class of deep neural network that is mostly used in the field of computer vision and imaging. CNNs are used to identify images, cluster them by their similarity, and implement object recognition within scenes. CNN has different layers— namely, the input layer, the output layer, and multiple hidden layers. These hidden layers of a CNN consist of fully connected layers, convolutional layers, a **ReLU layer** as an **activation function**, **normalization layers**, and **pooling layers**. On a very simple level, CNNs help us identify images and label them appropriately; for example, a tiger image will be identified as a tiger:

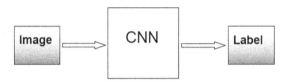

Figure 7.1: A generalized CNN

The following is an example of a CNN classifying a tiger:

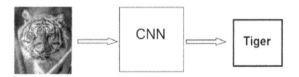

Figure 7.2: A CNN classifying an image of a tiger into the class "Tiger"

THE ARCHITECTURE OF A CNN

The main components of CNN architecture are as follows:

- Input image
- Convolutional layer
- Pooling layer
- Flattening

INPUT IMAGE

An **input image** forms the first component of a CNN architecture. An image can be of any type: a human, an animal, scenery, a medical X-ray image, and so on. Each image is converted into a mathematical matrix of zeros and ones. The following figure explains how a computer views an image of the letter **T**.

All the blocks that have a value of one represent the data, while the zeros represent blank space:

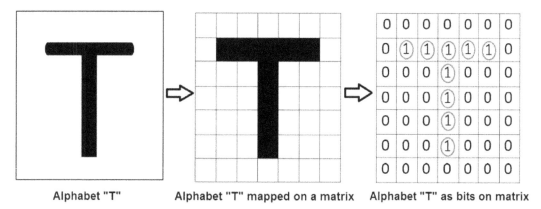

Alphabet "T"   Alphabet "T" mapped on a matrix   Alphabet "T" as bits on matrix

Figure 7.3: Matrix for the letter 'T'

CONVOLUTION LAYER

The **convolution layer** is the place where image processing starts. A convolution layer consists of two parts:

- **Feature detector** or **filter**
- **Feature map**

Feature detector or a **filter**: This is a matrix or pattern that you put on an image to transform it into a feature map:

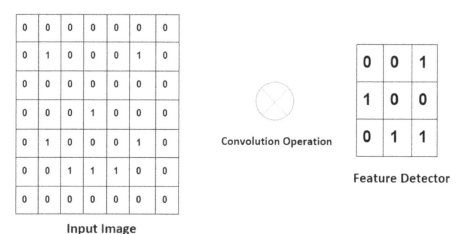

Input Image   Convolution Operation   Feature Detector

Figure 7.4: Feature detector

As we can see, this feature detector is put (superimposed) on the original image and the computation is done on the corresponding elements. The computation is done by multiplying the corresponding elements, as shown in the following figure. This process is repeated for all the cells. This results in a new processed image—
`(0x0+0x0+0x1) + (0x1+1x0+0x0) + (0x0+0x1+0x1) = 0`:

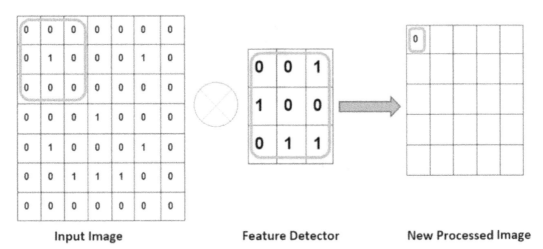

Input Image   Feature Detector   New Processed Image

Figure 7.5: Feature detector masked in an image

Feature Map: This is the reduced image that is produced by the convolution of an **image** and **feature detector**. We have to put the feature detector on all the possible locations of the original image and derive a smaller image from it; that derived image is the feature map of the input image:

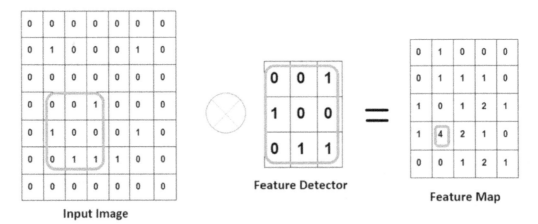

Input Image   Feature Detector   Feature Map

Figure 7.6: Feature map

> **NOTE**
>
> Here, the `feature detector` is the filter and the `feature map` is the reduced image. Some information is lost while reducing the image.

In an actual CNN, a number of feature detectors are used to produce a number of feature maps, as shown in the following figure:

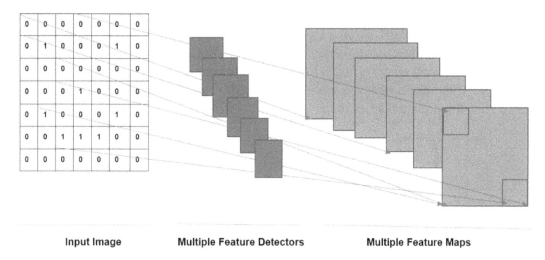

Input Image	**Multiple Feature Detectors**	**Multiple Feature Maps**

Figure 7.7: Multiple feature detectors and maps

THE POOLING LAYER

The `pooling layer` helps us ignore the less important data in the image and reduces the image further, all while preserving its important features. Consider the following three images, which contain four cats in total:

Figure 7.8: Example of cat images

To identify whether an image has a cat in it or not, the neural network analyzes the picture. It may look at ear shape, eye shape, and so on. At the same time, the image consists of lots of features that are not related to cats. The tree and leaves in the first two images are useless in the identification of the cat. The pooling mechanism helps the algorithm understand which parts of the image are relevant and which parts are irrelevant.

The feature map derived from the convolution layer is passed through a pooling layer to further reduce the image, all while preserving the most relevant part of the image. The pooling layer consists of functions such as max pooling, min pooling, and average pooling. What this means is that we select a matrix size, say **2x2**, and we scan the feature map and select the maximum number from the **2x2** matrix that fits in that block. The following image gives us a clear idea of how max pooling works. Refer to the colors; the max number in each of the colored boxes from the feature map is selected in the pooled feature map:

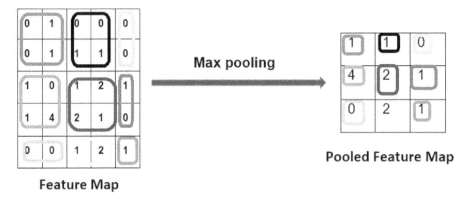

Figure 7.9: Pooling

Consider the case of the box that has number **4** in it. Let's assume that number **4** represents the ears of a cat, while the blank space around the ears is **0** and **1**. So, we ignore the **0** and **1** of that block and only select **4**. The following is some example code that we would use to add a pooling layer; here, **Maxpool2D** is used for max pooling, which helps identify the most important features:

```
classifier.add(MaxPool2D(2,2))
```

FLATTENING

Flattening is part of a CNN where the image is made ready to use as an input to an ANN. As the name suggests, the pooled image is flattened and converted into a single column. Each row is made into a column and stacked one over another. Here, we have converted a **3x3** matrix into a **1xn** matrix, where **n**, in our case, is **9**:

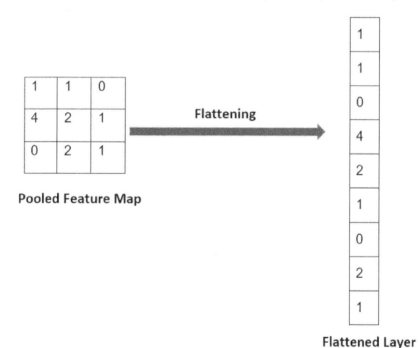

Figure 7.10: Flattening

In real-time, we have a number of pooled feature maps, and we flatten them into a single column. This single column is used as input for an ANN. The following figure shows a number of pooled layers flattened into a single column:

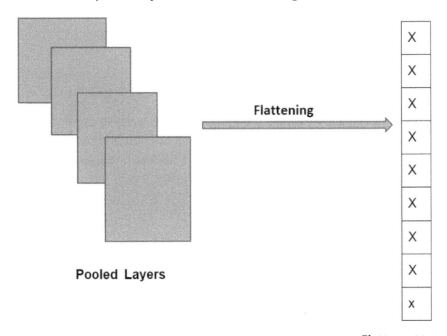

Figure 7.11: Pooling and flattening

The following is some example code that we would use to add a flattening layer; here **Flatten** is used for flattening the CNN:

```
classifier.add(Flatten())
```

Now, let's look at the overall structure of a CNN:

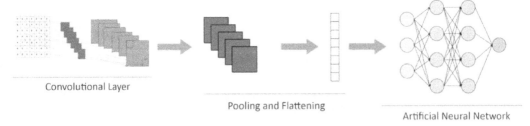

Figure 7.12: CNN architecture

The following is some example code that we would use to add the first layer to a CNN:

```
classifier.add(Conv2D(32,3,3,input_shape=(64,64,3),activation='relu'))
```

32,3,3 refers to the fact that there are **32** feature detectors of size **3x3**. As a good practice, always start with **32**; you can add **64** or **128** later.

Input_shape: Since all the images are of different shapes and sizes, this **input_image** converts all the images into a uniform shape and size. **(64,64)** is the dimension of the converted image. It can be set to **128** or **256**, but if you are working on a CPU on a laptop, it is advisable to use **64x64**. The last argument, **3**, is used because the image is a colored image (coded in red, blue, and green, or RGB). If the image is black and white, the argument can be set to one. The activation function that's being used is ReLU.

> **NOTE**
>
> We are using Keras with TensorFlow as the backend in this book. If the backend is Theano, then **input_image** will be coded as (**3,64,64**).

The last step is to fit the data that's been created. Here is the code that we use to do so:

```
classifier.fit_generator(training_set,steps_per_epoch = 5000,\
                         epochs = 25,validation_data = test_set,\
                         validation_steps = 1000)
```

> **NOTE**
>
> **steps_per_epoch** is the number of training images.
> **validation_steps** is the number of test images.

IMAGE AUGMENTATION

The word **augmentation** means the action or process of making or becoming greater in size or amount. **Image** or **data augmentation** works in a similar manner. Image/data augmentation creates many batches of our images. Then, it applies random transformations to random images inside the batches. Data transformation can be rotating images, shifting them, flipping them, and so on. By applying this transformation, we get more diverse images inside the batches, and we also have much more data than we had originally.

A cylinder can be rotated from different angles and seen differently. In the following figure, a single cylinder can be seen from five different angles. So, we have effectively created five different images from a single image:

Figure 7.13: Image augmentation of a cylinder

The following is some example code that we would use for image augmentation; here, the **ImageDataGenerator** class is used for processing. **shear_range**, **zoom_range**, and **horizontal_flip** are all used for transforming the images:

```
from keras.preprocessing.image import ImageDataGenerator

train_datagen = ImageDataGenerator(rescale = 1./255.0,\
                                   shear_range = 0.3,\
                                   zoom_range = 0.3,\
                                   horizontal_flip = False)
test_datagen = ImageDataGenerator(rescale = 1./255.0)
```

ADVANTAGES OF IMAGE AUGMENTATION

Image augmentation is an important part of processing images:

- **Reduces overfitting**: It helps reduce overfitting by creating multiple versions of the same image, rotated by a given amount.

- **Increases the number of images**: A single image acts as multiple images. So, essentially, the dataset has fewer images, but each image can be converted into multiple images with image augmentation. Image augmentation will increase the number of images and each image will be treated differently by the algorithm.

- **Easy to predict new images**: Imagine that a single image of a football is looked at from different angles and each angle is considered a distinct image. This will mean that the algorithm will be more accurate at predicting new images:

Figure 7.14: Image augmentation of an image of a football

Now that we have learned about the concepts and theory behind computer vision with CNNs, let's work on some practical examples.

First, we will start with an exercise in which we'll build a simple CNN. In the following exercises and activities, we will tweak our CNN using permutation and combining the following:

- Adding more CNN layers

- Adding more ANN layers

- Changing the optimizer function

- Changing the activation function

Let's begin by creating our first CNN so that we can classify images of cars and flowers into their respective classes.

EXERCISE 7.01: BUILDING A CNN AND IDENTIFYING IMAGES OF CARS AND FLOWERS

For this exercise, we have images of cars and flowers, which have been divided into training and testing sets, and we have to build a CNN that identifies whether an image is a car or a flower.

> **NOTE**
>
> All the exercises and activities in this chapter will be developed in Jupyter notebooks. Please download this book's GitHub repository, along with all the prepared templates, from https://packt.live/39tID2C.

Before you begin, ensure that you have downloaded the image datasets from this book's GitHub repository to your own working directory. You will need a **training_ set** folder to train your model and a **test_set** folder to test your model. Each of these folders will contain a **cars** folder, containing car images, and a **flowers** folder, containing flower images.

The steps for completing this exercise are as follows:

1. Import the **numpy** library and the necessary Keras libraries and classes:

```
# Import the Libraries
from keras.models import Sequential
from keras.layers import Conv2D, MaxPool2D, Flatten, Dense
import numpy as np
from tensorflow import random
```

2. Now, set a seed and initiate the model with the **Sequential** class:

```
# Initiate the classifier
seed = 1
np.random.seed(seed)
random.set_seed(seed)
classifier = Sequential()
```

3. Add the first layer of the **CNN**, set the input shape to **(64, 64, 3)**, the dimension of each image, and set the activation function as a **ReLU**:

```
classifier.add(Conv2D(32,3,3, input_shape=(64,64,3), \
            activation='relu'))
```

    32,3,3 shows that there are **32** feature detectors of **3x3** size.

4. Now, add the pooling layer with the image size as **2x2**:

```
classifier.add(MaxPool2D(2,2))
```

5. Flatten the output of the pooling layer by adding a flattening layer to the **CNN** model:

```
classifier.add(Flatten())
```

6. Add the first **Dense** layer of the **ANN**. Here, **128** is the output of the number of nodes. As a good practice, **128** is good to get started. **activation** is **relu**. As a good practice, the power of two is preferred:

```
classifier.add(Dense(128, activation='relu'))
```

7. Add the output layer of the ANN. This is a binary classification problem, so the size is **1** and the activation is **sigmoid**:

```
classifier.add(Dense(1, activation='sigmoid'))
```

8. Compile the network with an **adam** optimizer and compute the accuracy during the training process:

```
#Compile the network
classifier.compile(optimizer='adam', loss='binary_crossentropy', \
            metrics=['accuracy'])
```

9. Create training and test data generators. Rescale the training and test images by **1/255** so that all the values are between **0** and **1**. Set these parameters for the training data generators only – **shear_range=0.2**, **zoom_range=0.2**, and **horizontal_flip=True**:

```
from keras.preprocessing.image import ImageDataGenerator
train_datagen = ImageDataGenerator(rescale = 1./255,\
                                    shear_range = 0.2,\
                                    zoom_range = 0.2,\
                                    horizontal_flip = True)
test_datagen = ImageDataGenerator(rescale = 1./255)
```

10. Create a training set from the **training set** folder. **'../dataset/training_set'** is the folder where our data has been placed. Our CNN model has an image size of **64x64**, so the same size should be passed here too. **batch_size** is the number of images in a single batch, which is **32**. **Class_mode** is set to **binary** since we are working on binary classifiers:

```
training_set = train_datagen.flow_from_directory(\
               '../dataset/training_set',\
               target_size = (64, 64),\
               batch_size = 32,\
               class_mode = 'binary')
```

11. Repeat *step 10* for the test set while setting the folder to the location of the test images, that is, **'../dataset/test_set'**:

```
test_set = test_datagen.flow_from_directory(\
           '../dataset/test_set',\
           target_size = (64, 64),\
           batch_size = 32,\
           class_mode = 'binary')
```

12. Finally, fit the data. Set the **steps_per_epoch** to **10000** and the **validation_steps** to **2500**. The following step might take some time to execute:

```
classifier.fit_generator(training_set,steps_per_epoch = 10000,\
                         epochs = 2,validation_data = test_set,\
                         validation_steps = 2500,shuffle=False)
```

The preceding code produces the following output:

```
Epoch 1/2
10000/10000 [==============================] - 1994s 199ms/step -
loss: 0.2474 - accuracy: 0.8957 - val_loss: 1.1562 - val_accuracy:
0.8400
Epoch 2/2
10000/10000 [==============================] - 1695s 169ms/step -
loss: 0.0867 - accuracy: 0.9689 - val_loss: 1.4379 - val_accuracy:
0.8422
```

The accuracy on the validation set is **84.22%**.

> **NOTE**
>
> To get more accurate results, try increasing the number of epochs to about **25**. This will increase the time that it takes to process the data, and the total time is dependent on the configuration of your machine.
>
> To access the source code for this specific section, please refer to https://packt.live/38njqHU.
>
> You can also run this example online at https://packt.live/3iqFpSN.

That completes this exercise on processing images and identifying the contents of the images. An important thing to remember here is that this is a robust code for any binary classification problem in computer vision. This means that the code remains the same, even if the image data changes. We will test our knowledge of this by modifying some of the parameters of our model in the next activity and evaluating the model's performance.

ACTIVITY 7.01: AMENDING OUR MODEL WITH MULTIPLE LAYERS AND THE USE OF SOFTMAX

Since we have run a **CNN model** successfully, the next logical step is to try and improve the performance of our algorithm. There are many ways to improve its performance, and one of the most straightforward ways is by adding multiple ANN layers to the model, which we will learn about in this activity. We will also change the activation from sigmoid to softmax. By doing this, we can compare the result with that of the previous exercise. Follow these steps to complete this activity:

1. To build a CNN import library, set a seed and create a **Sequential** class and import **Conv2D, MaxPool2D, Flatten**, and **Dense**. **Conv2D** is used to build the convolution layer. Since our pictures are in 2D, we have used 2D here. Similarly, **Maxpool2D** is used for max pooling, **Flatten** is used for flattening the CNN, and **Dense** is used to add a fully connected CNN to an ANN.

2. Start building a CNN architecture using the preceding libraries. After adding the first layer, add two additional layers to your CNN.

3. Add a pooling and flattening layer to it, which will serve as the input for the ANN.

4. Build a fully connected ANN whose inputs will be the output of the CNN. After adding the first layer of your ANN, add three additional layers. For the output layer of your ANN, use the softmax activation function. Compile the model.

5. Perform image augmentation to process and transform the data. The **ImageDataGenerator** class is used for processing. **shear_range, zoom_range**, and **horizontal_flip** are all used for the transformation of images.

6. Create the training and test set data.

7. Lastly, fit the data that's been created.

After implementing these steps, you should get the following expected output:

```
Epoch 1/2
10000/10000 [==============================] - 2452s 245ms/step -
loss: 8.1783 - accuracy: 0.4667 - val_loss: 11.4999 - val_accuracy:
0.4695
Epoch 2/2
10000/10000 [==============================] - 2496s 250ms/step -
loss: 8.1726 - accuracy: 0.4671 - val_loss: 10.5416 - val_accuracy:
0.4691
```

> **NOTE**
>
> The solution for this activity can be found on page 439.

In this activity, we have modified our CNN model to try and improve the accuracy of our image classifier. We have added additional convolutional layers and additional ANN fully connected layers and changed the activation function in the output layer. By doing so our accuracy has decreased. In the next exercise, we will change the activation function back to a sigmoid. We will evaluate the performance by observing the accuracy evaluated on the validation dataset.

EXERCISE 7.02: AMENDING OUR MODEL BY REVERTING TO THE SIGMOID ACTIVATION FUNCTION

In this exercise, we will rebuild our model but revert the activation function from softmax back to sigmoid. By doing this, we can compare the accuracy with our previous model's. Follow these steps to complete this exercise:

1. Import the **numpy** library and the necessary Keras libraries and classes:

```
# Import the Libraries
from keras.models import Sequential
from keras.layers import Conv2D, MaxPool2D, Flatten, Dense
import numpy as np
from tensorflow import random
```

2. Now, set the seed and initiate the model with the **Sequential** class:

```
# Initiate the classifier
seed = 43
np.random.seed(seed)
random.set_seed(seed)
classifier = Sequential()
```

3. Add the first layer of the CNN, set the input shape to **(64, 64, 3)**, the dimension of each image, and set the activation function as a ReLU. Then, add **32** feature detectors of size **(3, 3)**. Add two additional convolutional layers with **32** feature detectors of size **(3, 3)**, also with ReLU activation functions:

```
classifier.add(Conv2D(32,3,3,input_shape=(64,64,3),\
                      activation='relu'))
classifier.add(Conv2D(32, (3, 3), activation = 'relu'))
classifier.add(Conv2D(32, (3, 3), activation = 'relu'))
```

4. Now, add the pooling layer with the image size as **2x2**:

```
classifier.add(MaxPool2D(2,2))
```

5. Add one more **Conv2D** with the same parameters as in *step 3* and a pooling layer to supplement it with the same parameters that we used in *step 4*:

```
classifier.add(Conv2D(32, (3, 3), activation = 'relu'))
classifier.add(MaxPool2D(pool_size = (2, 2)))
```

6. Flatten the output of the pooling layer by adding a flattening layer to the **CNN model**:

```
classifier.add(Flatten())
```

7. Add the first **Dense** layer of the ANN. Here, **128** is the output of the number of nodes. As a good practice, **128** is good to get started. **activation** is **relu**. As a good practice, the power of two is preferred. Add three additional layers with the same parameters:

```
classifier.add(Dense(128,activation='relu'))
classifier.add(Dense(128,activation='relu'))
classifier.add(Dense(128,activation='relu'))
classifier.add(Dense(128,activation='relu'))
```

8. Add the output layer of the **ANN**. This is a binary classification problem, so the output is **1** and the activation is **sigmoid**:

```
classifier.add(Dense(1,activation='sigmoid'))
```

9. Compile the network with an Adam optimizer and compute the accuracy during the training process:

```
classifier.compile(optimizer='adam', loss='binary_crossentropy', \
                   metrics=['accuracy'])
```

10. Create training and test data generators. Rescale the training and test images by **1/255** so that all the values are between **0** and **1**. Set these parameters for the training data generators only – **shear_range=0.2**, **zoom_range=0.2**, and **horizontal_flip=True**:

```
from keras.preprocessing.image import ImageDataGenerator
train_datagen = ImageDataGenerator(rescale = 1./255,
                                   shear_range = 0.2,
                                   zoom_range = 0.2,
                                   horizontal_flip = True)
test_datagen = ImageDataGenerator(rescale = 1./255)
```

11. Create a training set from the **training set** folder. **../dataset/ training_set** is the folder where our data is placed. Our CNN model has an image size of 64x64, so the same size should be passed here too. **batch_size** is the number of images in a single batch, which is **32**. **class_mode** is binary since we are working on binary classifiers:

```
training_set = \
train_datagen.flow_from_directory('../dataset/training_set',\
                                  target_size = (64, 64),\
                                  batch_size = 32,\
                                  class_mode = 'binary')
```

12. Repeat *step 11* for the test set by setting the folder to the location of the test images, that is, **'../dataset/test_set'**:

```
test_set = \
test_datagen.flow_from_directory('../dataset/test_set',\
                                 target_size = (64, 64),\
                                 batch_size = 32,\
                                 class_mode = 'binary')
```

13. Finally, fit the data. Set the **steps_per_epoch** to **10000** and the **validation_steps** to **2500**. The following step might take some time to execute:

```
classifier.fit_generator(training_set,steps_per_epoch = 10000,\
                         epochs = 2,validation_data = test_set,\
                         validation_steps = 2500,shuffle=False)
```

The preceding code produces the following output:

```
Epoch 1/2
10000/10000 [==============================] - 2241s 224ms/step -
loss: 0.2339 - accuracy: 0.9005 - val_loss: 0.8059 - val_accuracy:
0.8737
Epoch 2/2
10000/10000 [==============================] - 2394s 239ms/step -
loss: 0.0810 - accuracy: 0.9699 - val_loss: 0.6783 - val_accuracy:
0.8675
```

The accuracy of the model is **86.75%**, which is clearly greater than the accuracy of the model we built in the previous exercise. This shows the importance of activation functions. Just changing the output activation function from softmax to sigmoid increased the accuracy from **46.91%** to **86.75%**.

> **NOTE**
>
> To access the source code for this specific section, please refer to https://packt.live/2ZD9nKM.
>
> You can also run this example online at https://packt.live/3dPZiiQ.

In the next exercise, we will experiment with a different optimizer and observe how that affects the model's performance.

> **NOTE**
>
> In a binary classification problem (in our case, cars versus flowers), it is always better to use sigmoid as the activation function for the output.

EXERCISE 7.03: CHANGING THE OPTIMIZER FROM ADAM TO SGD

In this exercise, we will amend the model again by changing the optimizer to **SGD**. By doing this, we can compare the accuracy with our previous models. Follow these steps to complete this exercise:

1. Import the **numpy** library and the necessary Keras libraries and classes:

```
# Import the Libraries
from keras.models import Sequential
from keras.layers import Conv2D, MaxPool2D, Flatten, Dense
import numpy as np
from tensorflow import random
```

2. Now, initiate the model with the **Sequential** class:

```
# Initiate the classifier
seed = 42
np.random.seed(seed)
random.set_seed(seed)
classifier = Sequential()
```

3. Add the first layer of the **CNN**, set the input shape to **(64, 64, 3)**, the dimension of each image, and set the activation function as **ReLU**. Then, add **32** feature detectors of size (**3, 3**). Add two additional convolutional layers with the same number of feature detectors with the same size:

```
classifier.add(Conv2D(32,(3,3),input_shape=(64,64,3),\
                activation='relu'))
classifier.add(Conv2D(32,(3,3),activation='relu'))
classifier.add(Conv2D(32,(3,3),activation='relu'))
```

4. Now, add the pooling layer with the image size as **2x2**:

```
classifier.add(MaxPool2D(pool_size=(2, 2)))
```

5. Add one more **Conv2D** with the same parameters as in *step 3* and a pooling layer to supplement it with the same parameters that we used in *step 4*:

```
classifier.add(Conv2D(32, (3, 3), input_shape = (64, 64, 3), \
             activation = 'relu'))
classifier.add(MaxPool2D(pool_size=(2, 2)))
```

6. Add a **Flatten** layer to complete the CNN architecture:

```
classifier.add(Flatten())
```

7. Add the first **Dense** layer of the ANN of size **128**. Add three more dense layers to the network with the same parameters:

```
classifier.add(Dense(128,activation='relu'))
classifier.add(Dense(128,activation='relu'))
classifier.add(Dense(128,activation='relu'))
classifier.add(Dense(128,activation='relu'))
```

8. Add the output layer of the ANN. This is a binary classification problem, so the output is **1** and the activation is **sigmoid**:

```
classifier.add(Dense(1,activation='sigmoid'))
```

9. Compile the network with an **SGD optimizer** and compute the accuracy during the training process:

```
classifier.compile(optimizer='SGD', loss='binary_crossentropy', \
                  metrics=['accuracy'])
```

10. Create training and test data generators. Rescale the training and test images by **1/255** so that all the values are between **0** and **1**. Set these parameters for the training data generators only – **shear_range=0.2**, **zoom_range=0.2**, and **horizontal_flip=True**:

```
from keras.preprocessing.image import ImageDataGenerator

train_datagen = ImageDataGenerator(rescale = 1./255,\
                                   shear_range = 0.2,\
                                   zoom_range = 0.2,\
                                   horizontal_flip = True)
test_datagen = ImageDataGenerator(rescale = 1./255)
```

11. Create a training set from the **training set** folder. **../dataset/ training_set** is the folder where our data is placed. Our CNN model has an image size of **64x64**, so the same size should be passed here too. **batch_size** is the number of images in a single batch, which is **32**. **class_mode** is binary since we are creating a binary classifier:

```
training_set = \
train_datagen.flow_from_directory('../dataset/training_set',\
                                   target_size = (64, 64),\
                                   batch_size = 32,\
                                   class_mode = 'binary')
```

12. Repeat *step 11* for the test set by setting the folder to the location of the test images, that is, **'../dataset/test_set'**:

```
test_set = \
test_datagen.flow_from_directory('../dataset/test_set',\
                                  target_size = (64, 64),\
                                  batch_size = 32,\
                                  class_mode = 'binary')
```

13. Finally, fit the data. Set the **steps_per_epoch** to **10000** and the **validation_steps** to **2500**. The following step might take some time to execute:

```
classifier.fit_generator(training_set,steps_per_epoch = 10000,\
                         epochs = 2,validation_data = test_set,\
                         validation_steps = 2500,shuffle=False)
```

The preceding code produces the following output:

```
Epoch 1/2
10000/10000 [==============================] - 4376s 438ms/step -
loss: 0.3920 - accuracy: 0.8201 - val_loss: 0.3937 - val_accuracy:
0.8531

Epoch 2/2
10000/10000 [==============================] - 5146s 515ms/step -
loss: 0.2395 - accuracy: 0.8995 - val_loss: 0.4694 - val_accuracy:
0.8454
```

The accuracy is **84.54%** since we have used multiple **ANNs** and **SGD** as the optimizer.

> **NOTE**
>
> To access the source code for this specific section, please refer to https://packt.live/31Hu9vm.
>
> You can also run this example online at https://packt.live/3gqE9x8.

So far, we have worked with a number of different permutations and combinations of our model. It seems like the best accuracy for this dataset can be obtained by doing the following:

- Adding multiple CNN layers.

- Adding multiple ANN layers.

- Having the activation as sigmoid.

- Having the optimizer as adam.

- Increasing the epoch size to about **25** (this takes a lot of computational time – make sure you have a GPU to do this). This will increase the accuracy of your predictions.

Finally, we will go ahead and predict a new unknown image, pass it to the algorithm, and validate whether the image is classified correctly. In the next exercise, we will demonstrate how to use the model to classify new images.

EXERCISE 7.04: CLASSIFYING A NEW IMAGE

In this exercise, we will try to classify a new image. The image hasn't been exposed to the algorithm, so we will use this exercise to test our algorithm. You can run any of the algorithms in this chapter (although the one that gets the highest accuracy is preferred) and then use the model to classify the image.

> **NOTE**
>
> The image that's being used in this exercise can be found in this book's GitHub repository at https://packt.live/39tID2C.

Before we begin, ensure that you have downloaded **test_image_1** from this book's GitHub repository to your own working directory. This exercise follows on from the previous exercises, so ensure that you have one of the algorithms from this chapter ready to run in your workspace.

The steps for completing this exercise are as follows:

1. Load the image. **'test_image_1.jpg'** is the path of the test image. Please change the path to where you have saved the dataset in your system. Look at the image to verify what it is:

```
from keras.preprocessing import image
new_image = image.load_img('../test_image_1.jpg', \
                           target_size = (64, 64))
new_image
```

2. Print the class labels located in the **class_indices** attribute of the training set:

```
training_set.class_indices
```

3. Process the image:

```
new_image = image.img_to_array(new_image)
new_image = np.expand_dims(new_image, axis = 0)
```

4. Predict the new image:

```
result = classifier.predict(new_image)
```

5. The **prediction** method will output the image as **1** or **0**. To map **1** and **0** to **flower** or **car**, use the **class_indices** method with an **if...else** statement, as follows:

```
if result[0][0] == 1:
    prediction = 'It is a flower'
else:
    prediction = 'It is a car'

print(prediction)
```

The preceding code produces the following output:

```
It is a car
```

test_image_1 is the image of a car (you can see this by viewing the image for yourself) and was correctly predicted to be a car by the model.

In this exercise, we trained our model and then gave the model an image of a car. By doing this, we found out that the algorithm is classifying the image correctly. You can train the model on any type of an image by using the same process. For example, if you train the model with scans of lung infections and healthy lungs, then the model will be able to classify whether a new scan represents an infected lung or a healthy lung.

> **NOTE**
>
> To access the source code for this specific section, please refer to https://packt.live/31I6B9F.
>
> You can also run this example online at https://packt.live/2BzmEMx.

In the next activity, we will put our knowledge into practice by using a model that we trained in *Exercise 7.04, Classifying a New Image*.

ACTIVITY 7.02: CLASSIFYING A NEW IMAGE

In this activity, you will try to classify another new image, just like we did in the preceding exercise. The image is not exposed to the algorithm, so we will use this activity to test our algorithm. You can run any of the algorithms in this chapter (although the one that gets the highest accuracy is preferred) and then use the model to classify your images. The steps to implement this activity are as follows:

1. Run any one of the algorithms from this chapter.

2. Load the image (**test_image_2**) from your directory.

3. Process the image using the algorithm.

4. Predict the subject of the new image. You can view the image yourself to check whether the prediction is correct.

> **NOTE**
>
> The image that's being used in this activity can be found in this book's GitHub repository at https://packt.live/39tID2C.

Before starting, ensure you have downloaded **test_image_2** from this book's GitHub repository to your own working directory. This activity follows on directly from the previous exercises, so please ensure that you have one of the algorithms from this chapter ready to run in your workspace.

After implementing these steps, you should get the following expected output:

```
It is a flower
```

> **NOTE**
>
> The solution for this activity can be found on page 442.

In this activity, we trained the most performant model in this chapter when given the various parameters that were modified, including the optimizer and the activation function in the output layer according to the accuracy on the validation dataset. We tested the classifier on a test image and found it to be correct.

SUMMARY

In this chapter, we studied why we need computer vision and how it works. We learned why computer vision is one of the hottest fields in machine learning. Then, we worked with convolutional neural networks, learned about their architecture, and looked at how we can build CNNs in real-life applications. We also tried to improve our algorithms by adding more ANN and CNN layers and by changing the activation and optimizer functions. Finally, we tried out different activation functions and loss functions.

In the end, we were able to successfully classify new images of cars and flowers through the algorithm. Remember, the images of cars and flowers can be substituted with any other images, such as tigers and deer, or MRI scans of brains with and without a tumor. Any binary classification computer imaging problem can be solved with the same approach.

In the next chapter, we will study an even more efficient technique for working on computer vision, which is less time-consuming and easier to implement. The following chapter will teach us how to fine-tune pre-trained models for our own applications that will help create more accurate models that can be trained in faster times. The models that will be used are called VGG-16 and ResNet50 and are popular pre-trained models that are used to classify images.

8

TRANSFER LEARNING AND PRE-TRAINED MODELS

OVERVIEW

This chapter introduces the concept of pre-trained models and utilizing them for different applications from those for which they were trained, known as transfer learning. By the end of this chapter, you will be able to apply feature extraction to pre-trained models, exploit pre-trained models for image classification, and apply fine-tuning to pre-trained models to classify images of flowers and cars into their respective classes. We will see that this achieves the same task that we completed in the previous chapter but with greater accuracy and shorter training times.

INTRODUCTION

In the previous chapter, we learned how to create a **Convolutional Neural Network** (**CNN**) from scratch with Keras. We experimented with different architectures by adding more convolutional and Dense layers and changing the activation function. We compared the performance of each model by classifying images of cars and flowers into their respective classes and comparing their accuracies.

In real-world projects, however, you almost never code a convolutional neural network from scratch. You always tweak and train them as per the requirements. This chapter will introduce you to the important concepts of **transfer learning** and **pre-trained networks** (also known as **pre-trained models**), both of which are used in the industry.

We will use images and, rather than building a CNN from scratch, we will match these images on pre-trained models to try and classify them. We will also tweak our models to make them more flexible. The models we will use in this chapter are called **VGG16** and **ResNet50**, and we will discuss them later in this chapter. Before we start working on pre-trained models, we need to understand transfer learning.

PRE-TRAINED SETS AND TRANSFER LEARNING

Humans learn by experience. We apply the knowledge we gain in one situation to similar situations we face in the future. Suppose you want to learn how to drive an SUV. You have never driven an SUV; all you know is how to drive a small hatchback car.

The dimensions of the SUV are considerably larger than the hatchback, so navigating the SUV in traffic will surely be a challenge. Still, some basic systems (such as the clutch, accelerator, and brakes) remain similar to that of the hatchback. So, knowing how to drive a hatchback will surely be of great help to you when you are learning to drive the SUV. All the knowledge that you acquired while driving a hatchback can be used when you learn to drive a big SUV.

This is precisely what transfer learning is. By definition, transfer learning is a concept in machine learning in which we store and use the knowledge gained in one activity while learning another similar activity. The hatchback-SUV model fits this definition perfectly.

Suppose we want to know whether a picture is of a dog or a cat; here, we can have two approaches. One is building a deep learning model from scratch and then passing on the new pictures to the networks. Another option is to use a pre-trained deep learning neural network model that has already been built by using cats' and dogs' images, instead of creating a neural network from scratch.

Using the pre-trained model saves us computational time and resources. There can be some unforeseen advantages of using a pre-trained network. For example, almost all the pictures of dogs and cats will have some more objects in the picture, such as trees, the sky, and furniture. We can even use this pre-trained network to identify objects such as trees, the sky, and furniture.

So, a pre-trained network is a saved network (a neural network, in the case of deep learning) that was trained on a very large dataset, mostly on image classification problems. To work on a pre-trained network, we need to understand the concepts of feature extraction and fine-tuning.

FEATURE EXTRACTION

To understand feature extraction, we need to revisit the architecture of a convolutional neural network.

You may recall that the full architecture of a **CNN**, at a high level, consists of the following components:

- A **convolution layer**
- A **pooling and flattening layer**
- An **Artificial Neural Network** (**ANN**)

The following figure shows a complete CNN architecture:

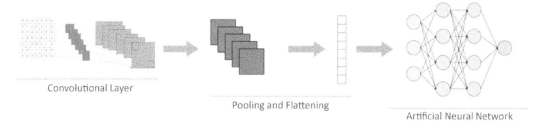

Figure 8.1: CNN architecture

Now, let's divide this architecture into two parts. The first part contains everything but the **ANN**, while the second part only contains the **ANN**. The following figure shows a split **CNN** architecture:

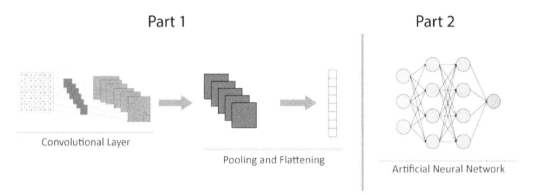

Figure 8.2: CNN split architecture – convolutional base and classifier

The first part is called a **convolutional base** while the second part is called the **classifier**.

In feature extraction, we keep reusing the convolutional base, and the classifier is changed. So, we preserve the learnings of the convolutional layer, and we can pass different classifiers on top of the convolutional layer. A classifier can be dog versus cat, bikes versus cars, or even medical X-ray images to classify tumors, infections, and so on. The following diagram shows some convolutional base layers that are used for different classifiers:

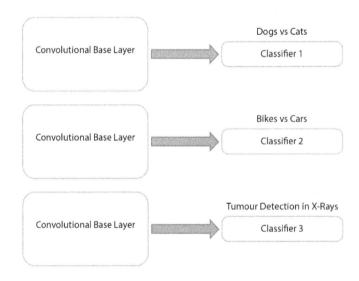

Figure 8.3: Reusable convolutional base layer

The obvious next question is, can't we reuse the classifier too, like the base layer? The general answer is no. The reason is that learning from the convolutional base is likely to be more generic and, therefore, more reusable. However, the learning of the classifier is mostly specific to the classes that the model was trained on. Therefore, it is advisable to only reuse the convolutional base layer and not the classifier.

The amount of generalized learning from a convolutional base layer depends on the depth of the layer. For example, in the case of a cat, the initial layers of the model learn about general traits such as edges and the background, while the higher layers may learn more about specific details such as eyes, ears, or the shape of the nose. So, if your new dataset is something very different from the original dataset—for example, if you wish to identify fruit instead of cats—then it is better to only use some initial layers of the convolutional base layer rather than using the whole layer.

Freezing convolutional layers: One of the most important features of pre-trained learning is to understand the concept of freezing some layers of a pre-trained network. Freezing essentially means that we stop the process of the weight updating some of the convolutional layers. Since we are using a pre-trained network, it is important to understand that we will need the information stored in the initial layers of the network. If that information is updated in training a network, we might lose some general concepts that have been learned and stored in the pre-trained network. If we add a classifier (**CNN**), many Dense layers on top of the network are randomly initialized, and there may be cases where, due to backpropagation, the learning of the initial layers of the network will be totally destroyed.

To avoid this information decay, we freeze some layers. This is done by making the layers non-trainable. The process of freezing some layers and training others is called fine-tuning a network.

FINE-TUNING A PRE-TRAINED NETWORK

Fine-tuning means tweaking our neural network in such a way that it becomes more relevant to the task at hand. We can freeze some of the initial layers of the network so that we don't lose information stored in those layers. The information stored there is generic and useful. However, if we can freeze those layers while our classifier is learning and then unfreeze them, we can tweak them a little so that they fit even better to the problem at hand. Suppose we have a pre-trained network that identifies animals. If we want to identify specific animals, such as dogs and cats, we can tweak the layers a little bit so that they can learn what dogs and cats look like. This is like using the whole pre-trained network and then adding a new layer that consists of images of dogs and cats. We will be doing a similar activity by using a pre-built network and adding a classifier on top of it, which will be trained on pictures of dogs and cats.

There is a three-point system to working with fine-tuning:

1. Add a classifier (**ANN**) on top of a pre-trained system.

2. Freeze the **convolutional base** and train the network.

3. Train the added **classifier** and the unfrozen part of the **convolutional base** jointly.

THE IMAGENET DATASET

In real practical work experience, you almost never need to build a base convolutional model on your own. You will always use pre-trained models. But where do you get the data from? For visual computing, the answer is ImageNet. The ImageNet dataset is a large visual database that is used in visual object recognition. It consists of more than 14 million labeled images with object names. ImageNet contains more than 20,000 categories.

SOME PRE-TRAINED NETWORKS IN KERAS

The following pre-trained networks can be thought of as the base convolutional layers. You use these networks and fit a classifier (ANN):

- `VGG16`

- `Inception V3`

- `Xception`

- `ResNet50`

- `MobileNet`

Different vendors have created the preceding pre-trained networks. For example, **ResNet50** was created by **Microsoft**, while **Inception V3** and **MobileNet** were created by **Google**. In this chapter, we will be working with the **VGG16** and **ResNet50** models.

VGG16 is a convolutional neural network model with 16 layers and was proposed by K. Simonyan and A. Zisserman from the University of Oxford. The model was submitted to the **ImageNet Large Scale Visual Recognition Challenge** (**ILSVRC**) in 2014—a challenge used to test state-of-the-art models that use the **ImageNet** dataset. **ResNet50** is another convolutional neural network that was trained on the **ImageNet** dataset that has 50 layers and won first place in the **ILSVRC** in 2015.

Now that we understand what these networks are, we will practice utilizing these pre-trained neural networks to classify an image of a slice of pizza with the **VGG16** model.

> ### NOTE
>
> All the exercises and activities in this chapter will be developed in Jupyter notebooks. Please download this book's GitHub repository, along with all the prepared templates, from https://packt.live/2uI63CC.

EXERCISE 8.01: IDENTIFYING AN IMAGE USING THE VGG16 NETWORK

We have a picture of a slice of pizza. We will use the **VGG16** network to process and identify the image. Before completing the following steps, ensure you have downloaded the **pizza** image from GitHub and saved it to your working directory:

1. Import the libraries:

```
import numpy as np
from keras.applications.vgg16 import VGG16
from keras.preprocessing import image
from keras.applications.vgg16 import preprocess_input
```

2. Initiate the model (this may take a while):

```
classifier = VGG16()
```

> **NOTE**
>
> The last layer of predictions (**Dense**) has 1,000 values. This means that **VGG16** has a total of 1,000 labels and our image will be one out of those 1,000 labels.

3. Load the image. `'../Data/Prediction/pizza.jpg.jpg'` is the path of the image on our system; it may be different on your system:

```
new_image= image.load_img('../Data/Prediction/pizza.jpg', \
                          target_size=(224, 224))
new_image
```

The following figure shows the output of the preceding code:

Figure 8.4: An image of a slice of pizza

The target size should be **224x224** since **VGG16** only accepts (**224,224**).

4. Change the image to an array by using the **img_to_array** function:

```
transformed_image = image.img_to_array(new_image)
transformed_image.shape
```

The preceding code produces the following output:

```
(224, 224, 3)
```

5. The image has to be in a four-dimensional form for **VGG16** to allow further processing. Expand the dimension of the image, as follows:

```
transformed_image = np.expand_dims(transformed_image, axis=0)
transformed_image.shape
```

The preceding code produces the following output:

```
(1, 224, 224, 3)
```

6. Preprocess the image using the **preprocess_input** function:

```
transformed_image = preprocess_input(transformed_image)
transformed_image
```

The following figure shows the output of the preceding code:

```
array([[[[ -65.939   ,   -83.779   ,   -91.68    ],
         [ -72.939   ,   -89.779   ,   -97.68    ],
         [ -73.939   ,   -88.779   ,   -95.68    ],
         ...,
         [-101.939   ,  -114.779   ,  -121.68    ],
         [ -89.939   ,  -103.779   ,  -108.68    ],
         [ -66.939   ,   -81.779   ,   -88.68    ]],

        [[ -66.939   ,   -83.779   ,   -91.68    ],
         [ -71.939   ,   -89.779   ,   -95.68    ],
         [ -66.939   ,   -81.779   ,   -89.68    ],
         ...,
         [ -98.939   ,  -113.779   ,  -121.68    ],
         [ -72.939   ,   -88.779   ,  -100.68    ],
         [ -49.939003,   -67.779   ,   -75.68    ]],
```

Figure 8.5: A screenshot of image preprocessing

7. Create the **predictor** variable:

```
y_pred = classifier.predict(transformed_image)
y_pred
```

8. Check the shape of the image. It should be (**1**, **1000**). It's **1000** because the **ImageNet** database has **1000** categories of images. The predictor variable shows the probability of our image being one of those images:

```
y_pred.shape
```

The preceding code produces the following output:

```
(1, 1000)
```

9. Print the top five probabilities of what our image is using the **decode_predictions** function and pass the function of the predictor variable, **y_pred**, and the number of predictions and corresponding labels to output:

```
from keras.applications.vgg16 import decode_predictions
decode_predictions(y_pred,top=5)
```

The preceding code produces the following output:

```
[[('n07873807', 'pizza', 0.97680503),
  ('n07871810', 'meat_loaf', 0.012848727),
  ('n07880968', 'burrito', 0.0019428912),
  ('n04270147', 'spatula', 0.0019108421),
  ('n03887697', 'paper_towel', 0.0009799759)]]
```

The first column of the array is the internal code number. The second is the possible label, while the third is the probability of the image being the label.

10. Put the predictions in a human-readable form. Print the most probable label from the output from the result of the **decode_predictions** function:

```
label = decode_predictions(y_pred)
"""
Most likely result is retrieved, for example, the highest probability
"""
decoded_label = label[0][0]
# The classification is printed
print('%s (%.2f%%)' % (decoded_label[1], \
        decoded_label[2]*100 ))
```

The preceding code produces the following output:

```
pizza (97.68%)
```

In this exercise, we predicted an image that says (with **97.68%** probability) that the picture is pizza. Clearly, higher accuracy here means a relatively similar object to our picture is present in the ImageNet database, and our algorithm has successfully identified the image.

> **NOTE**
>
> To access the source code for this specific section, please refer to https://packt.live/3dXqdsQ.
>
> You can also run this example online at https://packt.live/3dZMZAq.

In the following activity, we will put our knowledge to practice by using the **VGG16** network to classify an image of a motorbike.

ACTIVITY 8.01: USING THE VGG16 NETWORK TO TRAIN A DEEP LEARNING NETWORK TO IDENTIFY IMAGES

You are given an image of a motorbike. Use the **VGG16** network to predict the image. Before you start, ensure that you have downloaded the image (**test_image_1**) to your working directory. To complete this activity, follow these steps:

1. Import the required libraries, along with the **VGG16** network.

2. Initiate the pre-trained **VGG16** model.

3. Load the image that is going to be classified.

4. Preprocess the image by applying the transformations.

5. Create a predictor variable to predict the image.

6. Label the image and classify it.

> **NOTE**
>
> The solution for this activity can be found on page 444.

With that, we have completed this activity. Unlike in *Chapter 7, Computer Vision with Convolutional Neural Networks*, we did not build a **CNN** from scratch. Instead, we used a pre-trained model. We just uploaded a picture that needs to be classified. From this, we can see that, with **84.33%** accuracy, it is predicted to be a moped. In the next exercise, we'll work with an image for which there is no matching image in the ImageNet database.

EXERCISE 8.02: CLASSIFYING IMAGES THAT ARE NOT PRESENT IN THE IMAGENET DATABASE

Now, let's work with an image that is not part of the **1000** labels in our **VGG16** network. In this exercise, we will work with an image of a stick insect, and there are no labels for stick insects in our pre-trained network. Let's see what results we get:

1. Import the **numpy** library and the necessary **Keras** libraries:

```
import numpy as np
from keras.applications.vgg16 import VGG16
from keras.preprocessing import image
from keras.applications.vgg16 import preprocess_input
```

2. Initiate the model and print a summary of the model:

```
classifier = VGG16()
classifier.summary()
```

classifier.summary() shows us the architecture of the network. The following are the points to be noted – it has a four-dimensional input shape (**None, 224, 224, 3**) and it has three convolutional layers. The following figure shows the last four layers of the output:

```
flatten (Flatten)            (None, 25088)              0

fc1 (Dense)                  (None, 4096)               102764544

fc2 (Dense)                  (None, 4096)               16781312

predictions (Dense)          (None, 1000)               4097000
=================================================================
Total params: 138,357,544
Trainable params: 138,357,544
Non-trainable params: 0

None
```

Figure 8.6: Summary of the image using the VGG16 classifier

> **NOTE**
>
> The last layer of predictions (**Dense**) has **1000** values. This means that **VGG16** has a total of **1000** labels and that our image will be one out of those **1000** labels.

3. Load the image. **'../Data/Prediction/stick_insect.jpg'** is the path of the image on our system. It will be different on your system:

```
new_image = \
image.load_img('../Data/Prediction/stick_insect.jpg', \
               target_size=(224, 224))
new_image
```

The following figure shows the output of the preceding code:

Figure 8.7: Sample stick insect image for prediction

The target size should be **224x224** since **VGG16** only accepts (**224,224**).

4. Change the image to an array by using the **img_to_array** function:

```
transformed_image = image.img_to_array(new_image)
transformed_image.shape
```

5. The image must be in a four-dimensional form for **VGG16** to allow further processing. Expand the dimension of the image along the 0th axis using the **expand_dims** function:

```
transformed_image = np.expand_dims(transformed_image, axis=0)
transformed_image.shape
```

6. Preprocess the image using the **preprocess_input** function:

```
transformed_image = preprocess_input(transformed_image)
transformed_image
```

The following figure shows the output of the preceding code:

```
array([[[[-7.9390030e+00,  1.6221001e+01,  4.3320000e+01],
         [ 2.0609970e+00,  2.4221001e+01,  5.1320000e+01],
         [ 1.6060997e+01,  3.2221001e+01,  5.6320000e+01],
         ...,
         [ 7.0609970e+00,  2.1221001e+01,  4.0320000e+01],
         [-1.9390030e+00,  1.6221001e+01,  3.4320000e+01],
         [-6.9390030e+00,  1.4221001e+01,  3.1320000e+01]],

        [[ 9.0609970e+00,  3.3221001e+01,  6.0320000e+01],
         [ 6.0997009e-02,  2.2221001e+01,  4.9320000e+01],
         [ 8.0609970e+00,  3.1221001e+01,  5.4320000e+01],
         ...,
         [-6.9390030e+00,  1.1221001e+01,  2.9320000e+01],
         [-2.9390030e+00,  1.4221001e+01,  3.2320000e+01],
         [ 6.0609970e+00,  2.0221001e+01,  3.9320000e+01]],

        [[ 1.0060997e+01,  3.7221001e+01,  6.5320000e+01],
         [ 9.0609970e+00,  2.8221001e+01,  5.7320000e+01],
         [-9.3900299e-01,  1.8221001e+01,  4.6320000e+01],
         ...,
         [-6.9390030e+00,  9.2210007e+00,  2.9320000e+01],
         [-8.9390030e+00,  1.1221001e+01,  3.0320000e+01],
         [ 6.0997009e-02,  2.2221001e+01,  4.1320000e+01]],
```

Figure 8.8: Screenshot showing a few instances of image preprocessing

7. Create the **predictor** variable:

```
y_pred = classifier.predict(transformed_image)
y_pred
```

The following figure shows the output of the preceding code:

```
array([[4.21829981e-07, 1.85480451e-06, 1.72294085e-06, 6.76564525e-07,
        2.89053751e-05, 1.41852961e-05, 1.71890442e-05, 2.24749624e-06,
        3.92589482e-06, 3.78673963e-06, 2.06268323e-05, 7.51030393e-06,
        1.40643460e-05, 3.43733154e-05, 1.98462640e-05, 8.18990975e-06,
        1.08288223e-05, 1.76717931e-05, 1.64576650e-05, 3.33322532e-05,
        2.74088507e-05, 3.04659238e-06, 5.54778899e-06, 4.73525324e-06,
        3.29870386e-06, 1.77044087e-04, 1.83029479e-04, 5.83823072e-04,
        5.24099509e-04, 1.54459769e-06, 5.06804136e-05, 1.18027812e-04,
        2.67617492e-04, 1.81688793e-05, 3.93874470e-05, 4.16620605e-05,
        1.94424774e-05, 3.64137486e-05, 4.32395318e-04, 1.50895321e-05,
        6.48876361e-04, 8.92810232e-04, 4.67032398e-04, 1.95193552e-05,
        1.00129563e-03, 1.36421731e-04, 4.56671522e-04, 4.72387110e-05,
        5.16184491e-06, 2.23003917e-05, 1.38761870e-05, 3.23492600e-06,
        3.17189406e-04, 1.60120311e-04, 1.34436399e-04, 1.18729513e-05,
        1.03273931e-04, 1.06102932e-04, 9.10378076e-05, 3.55899065e-05,
        3.40783969e-04, 3.49663351e-05, 1.47626179e-05, 5.18944580e-06,
        2.36639626e-05, 2.91944925e-05, 1.46813414e-04, 8.76611521e-05,
        1.58484865e-04, 3.20514984e-04, 6.66521862e-02, 1.62668515e-03,
        5.70137578e-04, 3.67029905e-02, 4.12856275e-03, 1.79513693e-02,
```

Figure 8.9: Creating the predictor variable

8. Check the shape of the image. It should be **(1 , 1000)**. It's **1000** because, as we mentioned previously, the ImageNet database has **1000** categories of images. The predictor variable shows the probabilities of our image being one of those images:

```
y_pred.shape
```

The preceding code produces the following code:

```
(1, 1000)
```

9. Select the top five probabilities of what our image label is out of the **1000** labels that the **VGG16** network has:

```
from keras.applications.vgg16 import decode_predictions
decode_predictions(y_pred, top=5)
```

The preceding code produces the following code:

```
[[('n02231487', 'walking_stick', 0.30524516),
  ('n01775062', 'wolf_spider', 0.26035702),
  ('n03804744', 'nail', 0.14323168),
  ('n01770081', 'harvestman', 0.066652186),
  ('n01773549', 'barn_spider', 0.03670299)]]
```

The first column of the array is an internal code number. The second is the label, while the third is the probability of the image being the label.

10. Put the predictions in a human-readable format. Print the most probable label from the output from the result of the **decode_predictions** function:

```
label = decode_predictions(y_pred)
"""
Most likely result is retrieved, for example, the highest probability
"""
decoded_label = label[0][0]
# The classification is printed
print('%s (%.2f%%)' % (decoded_label[1], decoded_label[2]*100 ))
```

The preceding code produces the following code:

```
walking_stick (30.52%)
```

Here, you can see that the network predicted that our image was a walking stick with **30.52%** accuracy. Clearly, the image is not a walking stick but a stick insect; out of all the labels that the **VGG16** network contains, a walking stick is the closest thing to a stick insect. The following image is that of a walking stick:

Figure 8.10: Walking stick

To avoid such outputs, we could freeze the existing layer of **VGG16** and add our own layer. We could also add a layer that contains images of walking sticks and stick insects so that we can obtain better output.

If you have a large number of a walking stick and stick insect images, you could perform a similar task to improve the model's ability to classify images into their respective classes. You could then test it by rerunning the previous exercise.

> **NOTE**
>
> To access the source code for this specific section, please refer to https://packt.live/31I7bnR.
>
> You can also run this example online at https://packt.live/31Hv1QE.

To understand this in detail, let's work on a different example, where we freeze the last layer of the network and add our own layer with images of cars and flowers. This will help the network improve its accuracy in classifying images of cars and flowers.

EXERCISE 8.03: FINE-TUNING THE VGG16 MODEL

Let's work on fine-tuning the **VGG16** model. In this exercise, we will freeze the network and remove the last layer of **VGG16**, which has **1000** labels in it. After removing the last layer, we will build a new flower-car classifier **ANN**, just like we did in *Chapter 7, Computer Vision with Convolutional Neural Networks*, and will connect this **ANN** to **VGG16** instead of the original one with **1000** labels. Essentially, what we will do is replace the last layer of **VGG16** with a user-defined layer.

Before we begin, ensure you have downloaded the image datasets from this book's GitHub repository to your own working directory. You will need a `training_set` folder and a `test_set` folder to test your model. Each of these folders will contain a `cars` folder, containing car images, and a `flowers` folder, containing flower images.

The steps for completing this exercise are as follows:

> **NOTE**
>
> Unlike the original new model, which had **1000** labels (**100** different object categories), this new fine-tuned model will only have images of flowers or cars. So, whatever image you provide as an input to the model, it will categorize it as a flower or car based on its prediction probability.

1. Import the **numpy** library, TensorFlow's **random** library, and the necessary **Keras** libraries:

```
import numpy as np
import keras
from keras.layers import Dense
from tensorflow import random
```

2. Initiate the **VGG16** model:

```
vgg_model = keras.applications.vgg16.VGG16()
```

3. Check the model **summary**:

```
vgg_model.summary()
```

The following figure shows the output of the preceding code:

fc1 (Dense)	(None, 4096)	102764544
fc2 (Dense)	(None, 4096)	16781312
predictions (Dense)	(None, 1000)	4097000

```
=================================================================
Total params: 138,357,544
Trainable params: 138,357,544
Non-trainable params: 0
```

Figure 8.11: Model summary after initiating the model

4. Remove the last layer, **labeled predictions** in the preceding image, from the model summary. Create a new Keras model of the sequential class and iterate through all the layers of the VGG model. Add all of them to the new model, except for the last layer:

```
last_layer = str(vgg_model.layers[-1])

np.random.seed(42)
random.set_seed(42)
classifier= keras.Sequential()
for layer in vgg_model.layers:
    if str(layer) != last_layer:
        classifier.add(layer)
```

Here, we have created a new model name's classifier instead of **vgg_model**. All the layers, except the last layer, that is, **vgg_model**, have been included in the classifier.

5. Print the **summary** of the newly created model:

```
classifier.summary()
```

The following figure shows the output of the preceding code:

fc1 (Dense)	(None, 4096)	102764544
fc2 (Dense)	(None, 4096)	16781312

```
=================================================================
Total params: 134,260,544
Trainable params: 134,260,544
Non-trainable params: 0
```

Figure 8.12: Rechecking the summary after removing the last layer

The last layer of prediction (**Dense**) has been deleted.

6. Freeze the layers by iterating through the layers and setting the **trainable** parameter to **False**:

```
for layer in classifier.layers:
    layer.trainable=False
```

7. Add a new output layer of size **1** with a **sigmoid** activation function and print the model summary:

```
classifier.add(Dense(1, activation='sigmoid'))
classifier.summary()
```

The following function shows the output of the preceding code:

fc1 (Dense)	(None, 4096)	102764544
fc2 (Dense)	(None, 4096)	16781312
dense_1 (Dense)	(None, 1)	4097

```
=================================================================
Total params: 134,264,641
Trainable params: 4,097
Non-trainable params: 134,260,544
```

Figure 8.13: Rechecking the summary after adding the new layer

Now, the last layer is the newly created user-defined layer.

8. Compile the network with an **adam** optimizer and binary cross-entropy loss and compute the **accuracy** during training:

```
classifier.compile(optimizer='adam', loss='binary_crossentropy', \
                    metrics=['accuracy'])
```

Create some training and test data generators, just like we did in *Chapter 7, Computer Vision with Convolutional Neural Networks*. Rescale the training and test images by **1/255** so that all the values are between **0** and **1**. Set the following parameters for the training data generators only: **shear_range=0.2**, **zoom_range=0.2**, and **horizontal_flip=True**.

9. Next, create a training set from the **training set** folder. **../Data/dataset/training_set** is the folder where our data is placed. Our CNN model has an image size of **224x224**, so the same size should be passed here too. **batch_size** is the number of images in a single batch, which is **32**. **class_mode** is binary since we are creating a binary classifier.

> **NOTE**
>
> Unlike in *Chapter 7, Computer Vision with Convolutional Neural Networks*, where the image size was **64x64**, VGG16 needs an image size of **224x224**.

Finally, fit the model to the training data:

```python
from keras.preprocessing.image import ImageDataGenerator

generate_train_data = \
ImageDataGenerator(rescale = 1./255,\
                   shear_range = 0.2,\
                   zoom_range = 0.2,\
                   horizontal_flip = True)

generate_test_data = ImageDataGenerator(rescale =1./255)

training_dataset = \
generate_train_data.flow_from_directory(\
    '../Data/Dataset/training_set',\
    target_size = (224, 224),\
    batch_size = 32,\
    class_mode = 'binary')

test_datasetset = \
generate_test_data.flow_from_directory(\
    '../Data/Dataset/test_set',\
    target_size = (224, 224),\
    batch_size = 32,\
    class_mode = 'binary')

classifier.fit_generator(training_dataset,\
                         steps_per_epoch = 100,\
                         epochs = 10,\
                         validation_data = test_datasetset,\
                         validation_steps = 30,\
                         shuffle=False)
```

There are 100 training images here, so set **steps_per_epoch =100**, set **validation_steps=30**, and set **shuffle=False**:

```
100/100 [==============================] - 2083s 21s/step - loss:
0.5513 - acc: 0.7112 - val_loss: 0.3352 - val_acc: 0.8539
```

10. Predict the new image (the code is the same as it was in *Chapter 7, Computer Vision with Convolutional Neural Networks*). First, load the image from `'../Data/Prediction/test_image_2.jpg'` and set the target size to (**224, 224**) since the **VGG16** model accepts images of that size.

```
from keras.preprocessing import image
new_image = \
image.load_img('../Data/Prediction/test_image_2.jpg', \
               target_size = (224, 224))
new_image
```

At this point, you can view the image by executing the code **new_image** and the class labels by running **training_dataset.class_indices**.

Next, preprocess the image, first by converting the image into an array using the **img_to_array** function, then by adding another dimension along the 0th axis using the **expand_dims** function. Finally, make the prediction using the **predict** method of the classifier and printing the output in human-readable format:

```
new_image = image.img_to_array(new_image)
new_image = np.expand_dims(new_image, axis = 0)
result = classifier.predict(new_image)

if result[0][0] == 1:
    prediction = 'It is a flower'
else:
    prediction = 'It is a car'

print(prediction)
```

The preceding code produces the following output:

```
It is a car
```

11. As a final step, you can save the classifier by running **classifier.save('car-flower-classifier.h5')**.

Here, we can see that the algorithm has done the correct image classification by identifying the image of the car. We just used a pre-built **VGG16** model for image classification by tweaking its layers and molding it as per our requirements. This is a very powerful technique for image classification.

> **NOTE**
>
> To access the source code for this specific section, please refer to https://packt.live/2ZxCqzA
>
> This section does not currently have an online interactive example, and will need to be run locally.

In the next exercise, we will utilize a different pre-trained model, known as **ResNet50**, and demonstrate how to classify images with this model.

EXERCISE 8.04: IMAGE CLASSIFICATION WITH RESNET

Finally, before closing this chapter, let's work on an exercise with the **ResNet50** network. We'll use an image of a Nascar racer and try to predict it through the network. Follow these steps to complete this exercise:

1. Import the necessary libraries:

```
import numpy as np
from keras.applications.resnet50 import ResNet50, preprocess_input
from keras.preprocessing import image
```

2. Initiate the **ResNet50** model and print the **summary** of the model:

```
classifier = ResNet50()
classifier.summary()
```

The following figure shows the output of the preceding code:

```
activation_49 (Activation)       (None, 7, 7, 2048)   0      add_16[0][0]
_____
avg_pool (GlobalAveragePooling2 (None, 2048)          0      activation_49[0][0]
_____
fc1000 (Dense)                   (None, 1000)         2049000 avg_pool[0][0]
===============================================================================
Total params: 25,636,712
Trainable params: 25,583,592
Non-trainable params: 53,120
_____
None
```

Figure 8.14: A summary of the model

3. Load the image. `'../Data/Prediction/test_image_3.jpg'` is the path of the image on our system. It will be different on your system:

```
new_image = \
image.load_img('../Data/Prediction/test_image_3.jpg', \
          target_size=(224, 224))
new_image
```

The following figure shows the output of the preceding code:

Figure 8.15: Sample Nascar racer image for prediction

Note that the target size should be **224x224** since **ResNet50** only accepts **(224,224)**.

4. Change the image to an array by using the **img_to_array** function:

```
transformed_image = image.img_to_array(new_image)
transformed_image.shape
```

5. The image has to be in a four-dimensional form for **ResNet50** to allow further processing. Expand the dimension along the 0th axis using the **expand_dims** function:

```
transformed_image = np.expand_dims(transformed_image, axis=0)
transformed_image.shape
```

6. Preprocess the image using the **preprocess_input** function:

```
transformed_image = preprocess_input(transformed_image)
transformed_image
```

7. Create the predictor variable by using the classifier to predict the image using its **predict** method:

```
y_pred = classifier.predict(transformed_image)
y_pred
```

8. Check the shape of the image. It should be (**1 , 1000**):

```
y_pred.shape
```

The preceding code produces the following output:

```
(1, 1000)
```

9. Select the top five probabilities of what our image is using the **decode_predictions** function and by passing the predictor variable, **y_pred**, as the argument and the top number of predictions and corresponding labels:

```
from keras.applications.resnet50 import decode_predictions
decode_predictions(y_pred, top=5)
```

The preceding code produces the following output:

```
[[('n04037443', 'racer', 0.8013074),
  ('n04285008', 'sports_car', 0.06431753),
  ('n02974003', 'car_wheel', 0.024077434),
  ('n02504013', 'Indian_elephant', 0.019822922),
  ('n04461696', 'tow_truck', 0.007778575)]]
```

The first column of the array is an internal code number. The second is the label, while the third is the probability of the image being the label.

10. Put the predictions in a human-readable format. Print the most probable label from the output from the result of the **decode_predictions** function:

```
label = decode_predictions(y_pred)
"""
Most likely result is retrieved, for example, the highest probability
"""
decoded_label = label[0][0]
# The classification is printed
print('%s (%.2f%%)' % (decoded_label[1], \
      decoded_label[2]*100 ))
```

The preceding code produces the following output:

```
racer (80.13%)
```

Here, the model clearly shows (with a probability of **80.13%**) that the picture is that of a racer. This is the power of pre-trained models, and Keras gives us the flexibility to use and tweak these models.

> **NOTE**
>
> To access the source code for this specific section, please refer to https://packt.live/2BzvTMK.
>
> You can also run this example online at https://packt.live/3eWeljh.

In the next activity, we will classify another image using the pre-trained **ResNet50** model.

ACTIVITY 8.02: IMAGE CLASSIFICATION WITH RESNET

Now, let's work on an activity that uses another pre-trained network, known as **ResNet**. We have an image of television located at **../Data/Prediction/test_image_4**. We will use the **ResNet50** network to predict the image. To implement the activity, follow these steps:

1. Import the required libraries.

2. Initiate the **ResNet** model.

3. Load the image that needs to be classified.

4. Preprocess the image by applying the appropriate transformations.

5. Create a predictor variable to predict the image.

6. Label the image and classify it.

> **NOTE**
>
> The solution for this activity can be found on page 448.

So, the network says, with close to **100%** accuracy, that the image is that of a television. This time, we used a **ResNet50** pre-trained model to classify the image of television and obtained similar results to those we obtained using the **VGG16** model to predict the image of a slice of pizza.

SUMMARY

In this chapter, we covered the concept of transfer learning and how is it related to pre-trained networks. We utilized this knowledge by using the pre-trained deep learning networks **VGG16** and **ResNet50** to predict various images. We practiced how to take advantage of such pre-trained networks using techniques such as feature extraction and fine-tuning to train models faster and more accurately. Finally, we learned the powerful technique of tweaking existing models and making them work according to our dataset. This technique of building our own **ANN** over an existing **CNN** is one of the most powerful techniques used in the industry.

In the next chapter, we will learn about sequential modeling and sequential memory by looking at some real-life cases with Google Assistant. Furthermore, we will learn how sequential modeling is related to **Recurrent Neural Networks** (RNN). We will learn about the vanishing gradient problem in detail and how using an **LSTM** is better than a simple **RNN** to overcome the vanishing gradient problem. We will apply what we have learned to time series problems by predicting stock trends that come out as fairly accurate.

9

SEQUENTIAL MODELING WITH RECURRENT NEURAL NETWORKS

OVERVIEW

This chapter will introduce you to sequential modeling—creating models to predict the next value or series of values in a sequence. By the end of this chapter, you will be able to build sequential models, explain **Recurrent Neural Networks** (**RNNs**), describe the vanishing gradient problem, and implement **Long Short-Term Memory** (**LSTM**) architectures. You will apply **RNNs** with **LSTM** architectures to predict the value of the future stock price value of Alphabet and Amazon.

INTRODUCTION

In the previous chapter, we learned about pre-trained networks and how to utilize them for our own applications via transfer learning. We experimented with **VGG16** and **ResNet50**, two pre-trained networks that are used for image classification, and used them to classify new images and fine-tune them for our own applications. By utilizing pre-trained networks, we were able to train more accurate models quicker than the convolutional neural networks we trained in previous chapters.

In traditional neural networks (and every neural network architecture covered in prior chapters), data passes sequentially through the network from the input layer, and through the hidden layers (if any), to the output layer. Information passes through the network once and the outputs are considered independent of each other, and only dependent on the inputs to the model. However, there are instances where a particular output is dependent on the previous output of the system.

Consider the stock price of a company as an example: the output at the end of any given day is related to the output of the previous day. Similarly, in **Natural Language Processing** (**NLP**), the final words in a sentence are highly dependent on the previous words in the sentence if the sentence is to make grammatical sense. NLP is a specific application of sequential modeling in which the dataset being processed and analyzed is natural language data. A special type of neural network, called a **Recurrent Neural Network** (**RNN**), is used to solve these types of problems where the network needs to remember previous outputs.

This chapter introduces and explores the concepts and applications of RNNs. It also explains how RNNs are different from standard feedforward neural networks. You will also gain an understanding of the vanishing gradient problem and **Long-Short-Term-Memory** (**LSTM**) networks. This chapter also introduces you to sequential data and how it's processed. We will be working with share market data for stock price forecasting to learn all about these concepts.

SEQUENTIAL MEMORY AND SEQUENTIAL MODELING

If we analyze the stock price of **Alphabet** for the past 6 months, as shown in the following screenshot, we can see that there is a trend. To predict or forecast future stock prices, we need to gain an understanding of this trend and then do our mathematical computations while keeping this trend in mind:

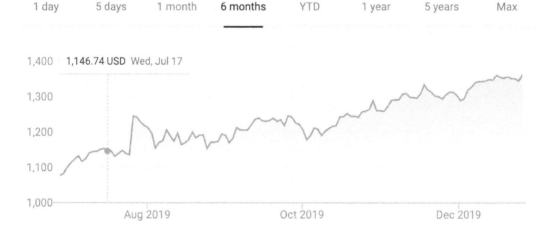

Figure 9.1: Alphabet's stock price over the last 6 months

This trend is deeply related to sequential memory and sequential modeling. If you have a model that can remember the previous outputs and then predict the next output based on the previous outputs, we say that the model has sequential memory.

The modeling that is done to process this sequential memory is known as **sequential modeling**. This is not only true for stock market data, but it is also true in NLP applications; we will look at one such example in the next section when we study RNNs.

RECURRENT NEURAL NETWORKS (RNNS)

RNNs are a class of neural networks that are built on the concept of sequential memory. Unlike traditional neural networks, an **RNN** predicts the results in sequential data. Currently, an **RNN** is the most robust technique that's available for processing sequential data.

If you have access to a smartphone that has Google Assistant, try opening it and asking the question: "When was the United Nations formed?" The answer is displayed in the following screenshot:

Figure 9.2: Google Assistant's output

Now, ask a second question, "Why was it formed?", as follows:

Here's a matching Wikipedia result

In 24 October 1945, at the end of World War II, the organization was established with the aim of preventing future wars. At its founding, the **UN** had 51 member states; there are now 193. The **UN** is the successor of the ineffective League of **Nations**.

United Nations - Wikipedia

Figure 9.3: Google Assistant's contextual output

Now, ask the third question, "Where are its headquarters?", and you should get the following answer:

Figure 9.4: Google Assistant's output

One interesting thing to note here is that we only mentioned the "United Nations" in the first question. In the second and third questions, we simply asked the assistant **why it was formed** and **where the headquarters were**, respectively. Google Assistant understood that since the previous question was about the United Nations, the next questions were also in the context of the United Nations. This is not a simple thing for a machine.

The machine was able to show the expected result because it had processed data in the form of a sequence. The machine understands that the current question is related to the previous question, and so, essentially, it remembers the previous question.

Let's consider another simple example. Say that we want to predict the next number in the following sequence: **7**, **8**, **9**, and **?**. We want the next output to be **9 + 1**. Alternatively, if we provide the sequence, **3**, **6**, **9**, and **?** we would like to get **9 + 3** as the output. While in both cases the last number is **9**, the prediction outcome should be different (that is, when we take into account the contextual information of the previous values and not only the last value). The key here is to remember the contextual information that was obtained from the previous values.

At a high level, such networks that can remember previous states are referred to as recurrent networks. To completely understand **RNNs**, let's revisit the traditional neural networks, also known as **feedforward neural networks**. This is a neural network in which the connections of the neural network do not form cycles; that is, the data only flows in one direction, as shown in the following diagram:

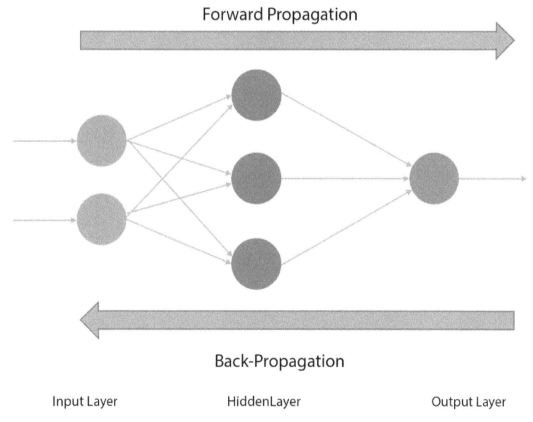

Figure 9.5: A feedforward neural network

In a **feedforward neural network**, such as the one shown in the preceding diagram, the input layer (the green circles on the left) gets the data and passes it to a hidden layer (with weights, illustrated by the blue circles in the middle). Later, the data from the hidden layer is passed to the output layer (illustrated by the red circle on the right). Based on the thresholds, the data is backpropagated, but there is no cyclical flow of data in the hidden layers.

In an **RNN**, the hidden layer of the network allows the cycle of data and information. As shown in the following diagram, the architecture is similar to a feedforward neural network; however, here, the data and information also flow in cycles:

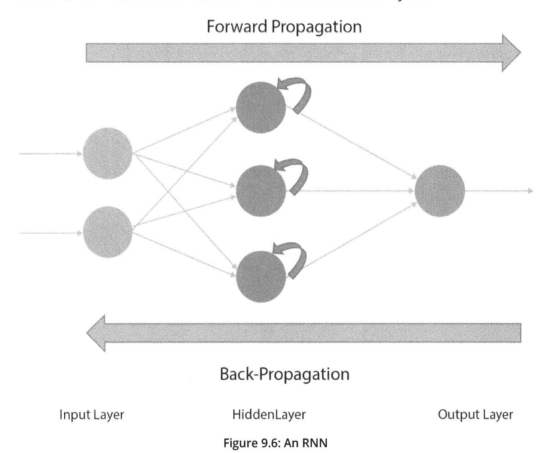

Figure 9.6: An RNN

Here, the defining property of the **RNN** is that the hidden layer not only gives the output, but it also feeds back the information of the output into itself. Before taking a deep dive into RNNs, let's discuss why we need **RNNs** and why `Convolutional Neural Networks (CNNs)` or normal `Artificial Neural Networks (ANNs)` fall short when it comes to processing sequential data. Suppose that we are using a **CNN** to identify images; first, we input an image of a dog, and the **CNN** will label the image as "dog". Then, we input an image of a mango, and the CNN will label the image as "mango". Let's input the image of the dog at time **t**, as follows:

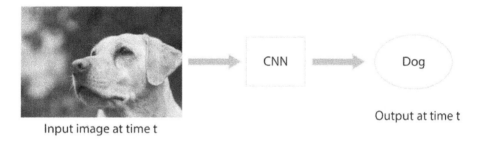

Input image at time t

Output at time t

Figure 9.7: An image of a dog with a CNN

Now, let's input the image of the mango at time **t + 1**, as follows:

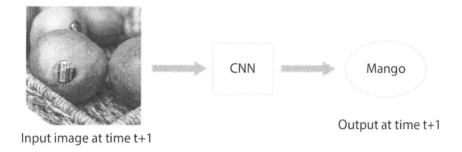

Input image at time t+1

Output at time t+1

Figure 9.8: An image of a mango with a CNN

Here, you can clearly see that the output at time **t** for the dog image and the output at time **t + 1** for the mango image are totally independent of each other. Therefore, we don't need our algorithms to remember previous instances of the output. However, as we mentioned in the Google Assistant example where we asked `when the United Nations was formed` and `why it was formed`, the output of the previous instance has to be remembered by the algorithm for it to process the sequential data. **CNNs** or **ANNs** are not able to do this, so we need to use **RNNs** instead.

In an **RNN**, we can have multiple outputs over multiple instances of time. The following diagram is a pictorial representation of an **RNN**. It represents the state of the network from time $t - 1$ to time $t + n$:

Output Layer

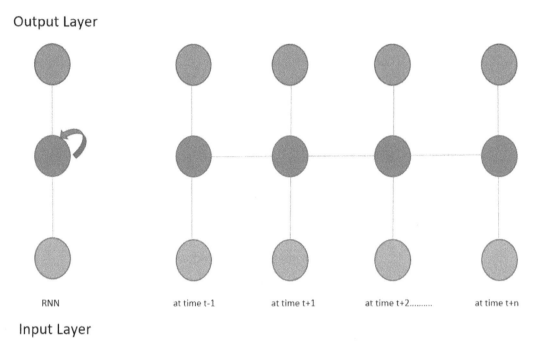

Input Layer

Figure 9.9: An unfolded RNN at various timestamps

There are some issues that you may face when training **RNNs** that are related to the unique architecture of **RNNs**. They concern the value of the gradient because, as the depth of the **RNN** increases, the gradient can either vanish or explode, as we will learn in the next section.

THE VANISHING GRADIENT PROBLEM

If someone asks you "What did you have for dinner last night?", it is pretty easy to remember and correctly answer them. Now, if someone asks you "What did you have for dinner over the past 30 days?", then you might be able to remember the menu of the past 3 or 4 days, but then the menu for the days before that will be a bit difficult to remember. This ability to recall information from the past is the basis of the vanishing gradient problem, which we will be studying in this section. Put simply, the vanishing gradient problem refers to information that is lost or has decayed over a period of time.

The following diagram represents the state of the **RNN** at different instances of time **t**. The top dots (in red) represent the output layer, the middle dots (in blue) represent the hidden layer, and the bottom dots (in green) represent the input layer:

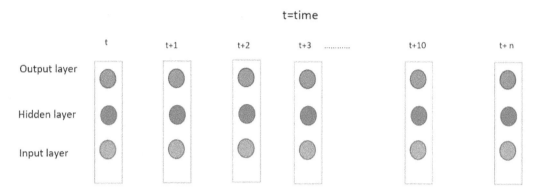

Figure: 9.10: Information decaying over time

If you are at **t + 10**, it will be difficult for you to remember what dinner menu you had at time **t** (which is 10 days prior to the current day). Additionally, if you are at **t + 100**, it is likely to be impossible for you to remember your dinner menu prior to 100 days, assuming that there is no pattern to the dinner you choose to make. In the context of machine learning, the vanishing gradient problem is a difficulty that is found when training ANNs using gradient-based learning methods and backpropagation. Let's recall how a neural network works, as follows:

1. First, we initialize the network with random weights and bias values.

2. We get a predicted output; this output is compared with the actual output and the difference is known as the cost.

3. The training process utilizes a gradient, which measures the rate at which the cost changes with respect to the weights or biases.

4. Then, we try to lower the cost by adjusting the weights and biases repeatedly throughout the training process, until the lowest possible value is obtained.

For example, if you place a ball on a steep floor, then the ball will roll quickly; however, if you place the ball on a flat surface, it will roll slowly, or not at all. Similarly, in a deep neural network, the model learns quickly when the gradient is large. However, if the gradient is small, then the model's learning rate becomes very low. Remember that, at any point, the gradient is the product of all the gradients up to that point (that is, it follows the calculus chain rule).

Additionally, the gradient is usually a small number between **0** and **1**, and the product of two numbers between **0** and **1** gives you an even smaller number. The deeper your network is, the smaller the gradient is in the initial layers of the network. In some cases, it reaches a point that is so small that no training happens in that network; this is the vanishing gradient problem. The following diagram shows the gradients following the calculus chain rule:

$$\frac{\partial C}{\partial b_1} = \sigma'(z_1) \times w_2 \times \sigma'(z_2) \times w_3 \times \sigma'(z_3) \times w_4 \times \sigma'(z_4) \times \frac{\partial C}{\partial a_4}$$

Figure 9.11: The vanishing gradient with cost, C, and the calculus chain rule

Referring to *Figure 9.10*, suppose that we are at the **t + 10** instance and we get an output that will be backpropagated to **t**, which is 10 steps away. Now, when the weight is updated, there will be 10 gradients (which are themselves very small), and when they multiply by each other, the number becomes so small that it is almost negligible. This is known as the vanishing gradient.

A BRIEF EXPLANATION OF THE EXPLODING GRADIENT PROBLEM

If instead of the weights being small, the weights are greater than **1**, then the subsequent multiplication will increase the gradient exponentially; this is known as the exploding gradient. The exploding gradient is simply the opposite of the vanishing gradient as in the case of the vanishing gradient, the values become too small, while in the case of the exploding gradient, the values become very large. As a result, the network suffers heavily and is unable to predict anything. We don't get the exploding gradient problem as frequently as vanishing gradients, but it is good to have a brief understanding of what exploding gradients are.

There are some approaches we take to overcome the challenges that are faced with the vanishing or exploding gradient problem. The one approach that we will learn about is **Long Short-Term Memory**, which overcomes issues with the gradients by having memory about information for long periods of time.

LONG SHORT-TERM MEMORY (LSTM)

LSTMs are **RNNs** whose main objective is to overcome the shortcomings of the vanishing gradient and exploding gradient problems. The architecture is built so that they remember data and information for a long period of time.

LSTMs were designed to overcome the limitation of the vanishing and exploding gradient problems. **LSTM** networks are a special kind of **RNN** that are capable of learning long-term dependencies. They are designed to avoid the long-term dependency problem; being able to remember information for long intervals of time is how they are wired. The following diagram displays a standard recurrent network where the repeating module has a **tanh activation** function. This is a simple **RNN**. In this architecture, we often have to face the vanishing gradient problem:

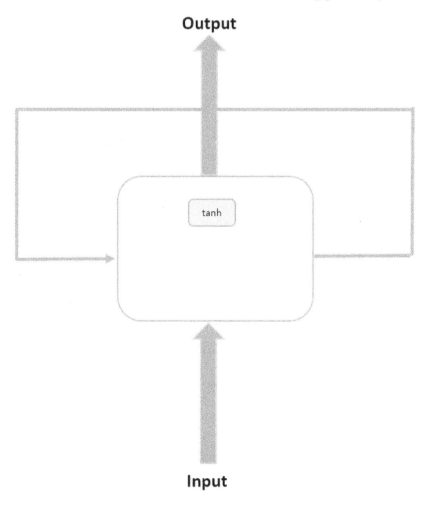

Figure 9.12: A simple RNN model

The **LSTM** architecture is similar to simple **RNNs**, but their repeating module has different components, as shown in the following diagram:

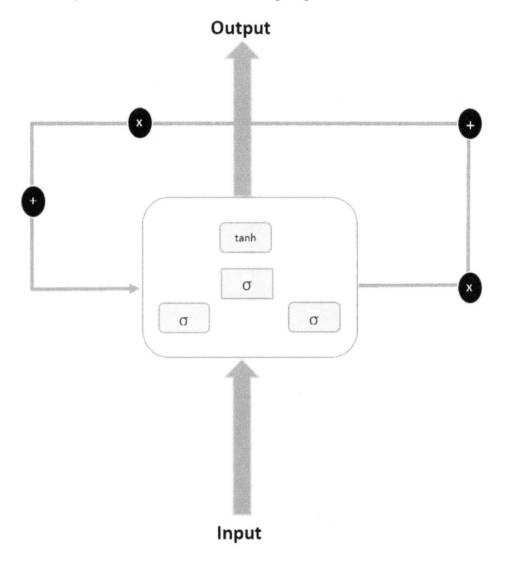

Figure 9.13: The LSTM model architecture

In addition to a simple **RNN**, an **LSTM** consists of the following:

- **Sigmoid activation** functions (σ)

- Mathematical computational functions (the black circles with + and x)

- Gated cells (or gates):

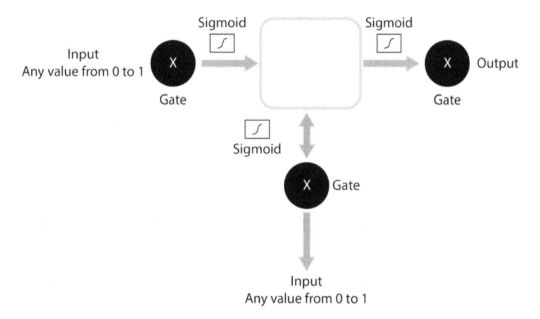

Figure 9.14: An LSTM in detail

The main difference between a simple **RNN** and an **LSTM** is the presence of gated cells. You can think of gates as computer memory, where information can be written, read, or stored. The preceding diagram shows a detailed image of an **LSTM**. The cells in the gates (represented by the black circles) make decisions on what to store and when to allow values to be read or written. The gates accept any information from **0** to **1**; that is, if it is 0, then the information is blocked; if it is **1**, then all the information flows through. If the input is between **0** and **1**, then only partial information flows.

Besides these input gates, the gradient of a network is dependent on two factors: weight and the activation function. The gates decide which piece of information needs to persist within the **LSTM** cell and which needs to be forgotten or deleted. In this way, the gates are like water valves; that is, the network can select which valve will allow the water to flow and which valve won't allow the water to flow.

The valves are adjusted in such a way that the values of the output will never yield a gradient (vanishing or exploding) problem. For example, if the value becomes too large, then there is a forget gate that will forget the value and no longer consider it for computations. Essentially, what a forget gate does is multiply the information by **0** or **1**. If the information needs to be processed further, then the forget gate multiplies the information by **1**, and if it needs to be forgotten, then it multiplies the information by **0**. Each gate is assisted by a sigmoid function that squashes the information between **0** and **1**. For us to gain a better understanding of this, let's take a look at some activities and exercises.

> **NOTE**
>
> All the activities and exercises in this chapter will be developed in Jupyter notebooks. You can download this book's GitHub repository, along with all the prepared templates, at https://packt.live/2vtdA8o.

EXERCISE 9.01: PREDICTING THE TREND OF ALPHABET'S STOCK PRICE USING AN LSTM WITH 50 UNITS (NEURONS)

In this exercise, we will examine the stock price of Alphabet over a period of 5 years— that is, from January 1, 2014, to December 31, 2018. In doing so, we will try to predict and forecast the company's future trend for January 2019 using **RNNs**. We have the actual values for January 2019, so we will be able to compare our predictions with the actual values later. Follow these steps to complete this exercise:

1. Import the required libraries:

```
import numpy as np
import matplotlib.pyplot as plt
import pandas as pd
from tensorflow import random
```

2. Import the dataset using the pandas **read_csv** function and look at the first five rows of the dataset using the **head** method:

```
dataset_training = pd.read_csv('../GOOG_train.csv')
dataset_training.head()
```

The following figure shows the output of the preceding code:

	Date	Open	High	Low	Close	Adj Close	Volume
0	2014-01-02	555.647278	556.788025	552.060730	554.481689	554.481689	3656400
1	2014-01-03	555.418152	556.379578	550.401978	550.436829	550.436829	3345800
2	2014-01-06	554.426880	557.340942	551.154114	556.573853	556.573853	3551800
3	2014-01-07	560.399475	567.717041	558.486633	567.303589	567.303589	5124300
4	2014-01-08	570.860291	571.517822	564.528992	568.484192	568.484192	4501700

Figure 9.15: The first five rows of the GOOG_Training dataset

3. We are going to make our prediction using the **Open** stock price; therefore, select the **Open** stock price column from the dataset and print the values:

```
training_data = dataset_training[['Open']].values
training_data
```

The preceding code produces the following output:

```
array([[ 555.647278],
       [ 555.418152],
       [ 554.42688 ],
       ...,
       [1017.150024],
       [1049.619995],
       [1050.959961]])
```

4. Then, perform feature scaling by normalizing the data using **MinMaxScaler** and setting the range of the features so that they have a minimum value of **0** and a maximum value of one. Use the **fit_transform** method of the scaler on the training data:

```
from sklearn.preprocessing import MinMaxScaler
sc = MinMaxScaler(feature_range = (0, 1))
training_data_scaled = sc.fit_transform(training_data)
training_data_scaled
```

The preceding code produces the following output:

```
array([[0.08017394],
       [0.07987932],
       [0.07860471],
       ...,
       [0.67359064],
       [0.71534169],
       [0.71706467]])
```

5. Create the data to get 60 timestamps from the current instance. We chose **60** here as this will give us a sufficient number of previous instances so that we can understand the trend; technically, this can be any number, but **60** is the optimal value. Additionally, the upper bound value here is **1258**, which is the index or count of rows (or records) in the training set:

```
X_train = []
y_train = []
for i in range(60, 1258):
    X_train.append(training_data_scaled[i-60:i, 0])
    y_train.append(training_data_scaled[i, 0])
X_train, y_train = np.array(X_train), \
                   np.array(y_train)
```

6. Next, reshape the data to add an extra dimension to the end of **X_train** using NumPy's **reshape** function:

```
X_train = np.reshape(X_train, (X_train.shape[0], \
                               X_train.shape[1], 1))
X_train
```

The following figure shows the output of the preceding code:

```
array([[[0.08017394],
        [0.07987932],
        [0.07860471],
        ...,
        [0.10999004],
        [0.09405781],
        [0.08533808]],

       [[0.07987932],
        [0.07860471],
        [0.08628449],
        ...,
        [0.09405781],
        [0.08533808],
        [0.09263448]],

       [[0.07860471],
        [0.08628449],
        [0.09973538],
        ...,
        [0.08533808],
        [0.09263448],
        [0.08214508]],
```

Figure 9.16: The data of a few timestamps from the current instance

7. Import the following Keras libraries to build the **RNN**:

```
from keras.models import Sequential
from keras.layers import Dense, LSTM, Dropout
```

8. Set the seed and initiate the sequential model, as follows:

```
seed = 1
np.random.seed(seed)
random.set_seed(seed)
model = Sequential()
```

9. Add an LSTM layer to the network with 50 units, set the **return_sequences** argument to **True**, and set the **input_shape** argument to **(X_train. shape[1], 1)**. Add three additional LSTM layers, each with 50 units, and set the **return_sequences** argument to **True** for the first two, as follows:

```
model.add(LSTM(units = 50, return_sequences = True, \
               input_shape = (X_train.shape[1], 1)))
# Adding a second LSTM layer
model.add(LSTM(units = 50, return_sequences = True))
# Adding a third LSTM layer
model.add(LSTM(units = 50, return_sequences = True))
# Adding a fourth LSTM layer
model.add(LSTM(units = 50))
# Adding the output layer
model.add(Dense(units = 1))
```

10. Compile the network with an **adam** optimizer and use **Mean Squared Error** for the loss. Fit the model to the training data for **100** epochs with a batch size of **32**:

```
# Compiling the RNN
model.compile(optimizer = 'adam', loss = 'mean_squared_error')


# Fitting the RNN to the Training set
model.fit(X_train, y_train, epochs = 100, batch_size = 32)
```

11. Load and process the **test** data (which is treated as actual data here) and select the column representing the value of **Open** stock data:

```
dataset_testing = pd.read_csv("../GOOG_test.csv")
actual_stock_price = dataset_testing[['Open']].values
actual_stock_price
```

The following figure shows the output of the preceding code:

```
array([[1016.570007],
       [1041.        ],
       [1032.589966],
       [1071.5       ],
       [1076.109985],
       [1081.650024],
       [1067.660034],
       [1063.180054],
       [1046.920044],
       [1050.170044],
       [1080.        ],
       [1079.469971],
       [1100.        ],
       [1088.        ],
       [1077.349976],
       [1076.47998 ],
       [1085.        ],
       [1080.109985],
       [1072.680054],
       [1068.430054],
       [1103.        ]])
```

Figure 9.17: The actual processed data

12. Concatenate the data; we will need **60** previous instances in order to get the stock price for each day. Therefore, we will need both training and test data:

```
total_data = pd.concat((dataset_training['Open'], \
                        dataset_testing['Open']), axis = 0)
```

13. Reshape and scale the input to prepare the test data. Note that we are predicting the January monthly trend, which has **21** financial days, so in order to prepare the test set, we take the lower bound value as 60 and the upper bound value as 81. This ensures that the difference of **21** is maintained:

```
inputs = total_data[len(total_data) \
         - len(dataset_testing) - 60:].values
inputs = inputs.reshape(-1,1)
inputs = sc.transform(inputs)
X_test = []
for i in range(60, 81):
    X_test.append(inputs[i-60:i, 0])
X_test = np.array(X_test)
X_test = np.reshape(X_test, (X_test.shape[0], \
                    X_test.shape[1], 1))
predicted_stock_price = model.predict(X_test)
predicted_stock_price = sc.inverse_transform(\
                    predicted_stock_price)
```

14. Visualize the results by plotting the actual stock price and then plotting the predicted stock price:

```
# Visualizing the results
plt.plot(actual_stock_price, color = 'green', \
         label = 'Real Alphabet Stock Price',\
         ls='--')
plt.plot(predicted_stock_price, color = 'red', \
         label = 'Predicted Alphabet Stock Price',\
         ls='-')
plt.title('Predicted Stock Price')
plt.xlabel('Time in days')
plt.ylabel('Real Stock Price')
plt.legend()
plt.show()
```

Please note that your results may differ slightly to the actual stock price of Alphabet.

Expected output:

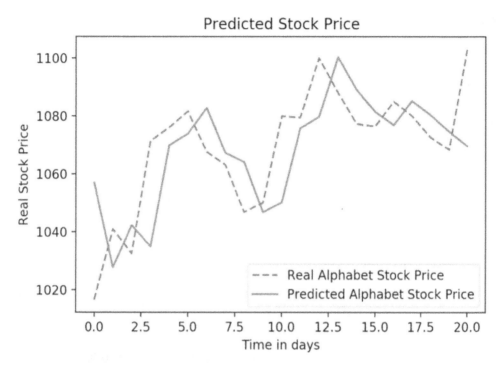

Figure 9.18: The real versus predicted stock price

This concludes *Exercise 9.01, Predicting the Trend of Alphabet's Stock Price Using an LSTM with 50 Units (Neurons)*, where we have predicted Alphabet's stock trends with the help of an **LSTM**. As shown in the preceding plot, the trend has been captured fairly.

> **NOTE**
>
> To access the source code for this specific section, please refer to https://packt.live/2ZwdAzW.
>
> You can also run this example online at https://packt.live/2YV3PvX.

In the next activity, we will test our knowledge and practice building **RNNs** with **LSTM** layers by predicting the trend of Amazon's stock price over the last 5 years.

ACTIVITY 9.01: PREDICTING THE TREND OF AMAZON'S STOCK PRICE USING AN LSTM WITH 50 UNITS (NEURONS)

In this activity, we will examine the stock price of Amazon for the last 5 years—that is, from January 1, 2014, to December 31, 2018. In doing so, we will try to predict and forecast the company's future trend for January 2019 using an RNN and LSTM. We have the actual values for January 2019, so we can compare our predictions to the actual values later. Follow these steps to complete this activity:

1. Import the required libraries.

2. From the full dataset, extract the **Open** column as the predictions will be made on the open stock value. Download the dataset from this book's GitHub repository. You can find the dataset at https://packt.live/2vtdA8o.

3. Normalize the data between 0 and 1.

4. Then, create timestamps. The values of each day in January 2019 will be predicted by the previous 60 days; so, if January 1 is predicted by using the value from the nth day up to December 31, then January 2 will be predicted by using the $n + 1^{st}$ day and January 1, and so on.

5. Reshape the data into three dimensions since the network needs data in three dimensions.

6. Build an **RNN** model in **Keras** using **50** units (here, units refer to neurons) with four **LSTM** layers. The first step should provide the input shape. Note that the final **LSTM** layer always adds **return_sequences=True**, so it doesn't have to be explicitly defined.

7. Process and prepare the test data that is the actual data for January 2019.

8. Combine and process the training and test data.

9. Visualize the results.

After implementing these steps, you should see the following expected output:

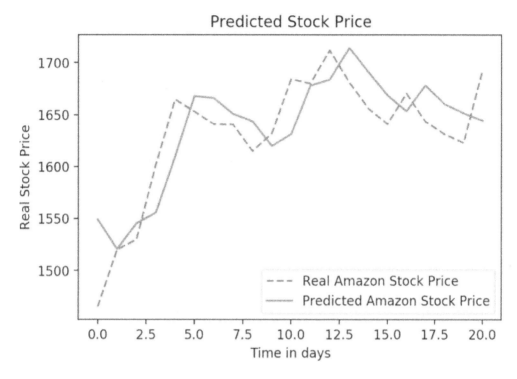

Figure 9.19: Real versus predicted stock prices

> **NOTE**
>
> The solution for this activity can be found on page 452.

Now, let's try and improve performance by tweaking our **LSTM**. There is no gold standard on how to build an **LSTM**; however, the following permutation combinations can be tried in order to improve performance:

- Build an **LSTM** with moderate units, such as **50**

- Build an **LSTM** with over **100** units

- Use more data; that is, instead of **5** years, take data from **10** years

- Apply regularization using **100** units

- Apply regularization using **50** units

- Apply regularization using more data and **50** units

This list can have a number of combinations; whichever combination offers the best results can be considered a good algorithm for that particular dataset. In the next exercise, we will explore one of these options by adding more units to our **LSTM** layer and observing the performance.

EXERCISE 9.02: PREDICTING THE TREND OF ALPHABET'S STOCK PRICE USING AN LSTM WITH 100 UNITS

In this exercise, we will examine the stock price of Alphabet over the last 5 years, from January 1, 2014, to December 31, 2018. In doing so, we will try to predict and forecast the company's future trend for January 2019 using RNNs. We have the actual values for January 2019, so we will compare our predictions with the actual values later. This is the same task as the first exercise, but now we're using 100 units instead. Make sure that you compare the output with *Exercise 9.01*, *Predicting the Trend of Alphabet's Stock Price Using an LSTM with 50 Units (Neurons)*. Follow these steps to complete this exercise:

1. Import the required libraries:

```
import numpy as np
import matplotlib.pyplot as plt
import pandas as pd
from tensorflow import random
```

2. Import the dataset using the pandas **read_csv** function and look at the first five rows of the dataset using the **head** method:

```
dataset_training = pd.read_csv('../GOOG_train.csv')
dataset_training.head()
```

3. We are going to make our prediction using the **Open** stock price; therefore, select the **Open** stock price column from the dataset and print the values:

```
training_data = dataset_training[['Open']].values
training_data
```

4. Then, perform feature scaling by normalizing the data using **MinMaxScaler** and setting the range of the features so that they have a minimum value of zero and a maximum value of one. Use the **fit_transform** method of the scaler on the training data:

```
from sklearn.preprocessing import MinMaxScaler
sc = MinMaxScaler(feature_range = (0, 1))
training_data_scaled = sc.fit_transform(training_data)
training_data_scaled
```

5. Create the data to get **60** timestamps from the current instance. We chose **60** here as it will give us a sufficient number of previous instances in order to understand the trend; technically, this can be any number, but **60** is the optimal value. Additionally, the upper bound value here is **1258**, which is the index or count of rows (or records) in the **training** set:

```
X_train = []
y_train = []
for i in range(60, 1258):
    X_train.append(training_data_scaled[i-60:i, 0])
    y_train.append(training_data_scaled[i, 0])
X_train, y_train = np.array(X_train), np.array(y_train)
```

6. Reshape the data to add an extra dimension to the end of **X_train** using NumPy's **reshape** function:

```
X_train = np.reshape(X_train, (X_train.shape[0], \
                               X_train.shape[1], 1))
```

7. Import the following **Keras** libraries to build the **RNN**:

```
from keras.models import Sequential
from keras.layers import Dense, LSTM, Dropout
```

8. Set the seed and initiate the sequential model, as follows:

```
seed = 1
np.random.seed(seed)
random.set_seed(seed)
model = Sequential()
```

9. Add an **LSTM** layer to the network with **50** units, set the **return_sequences** argument to **True**, and set the **input_shape** argument to **(X_train. shape[1], 1)**. Add three additional **LSTM** layers, each with **50** units, and set the **return_sequences** argument to **True** for the first two. Add a final output layer of size **1**:

```
model.add(LSTM(units = 100, return_sequences = True, \
               input_shape = (X_train.shape[1], 1)))

# Adding a second LSTM
model.add(LSTM(units = 100, return_sequences = True))

# Adding a third LSTM layer
model.add(LSTM(units = 100, return_sequences = True))

# Adding a fourth LSTM layer
model.add(LSTM(units = 100))

# Adding the output layer
model.add(Dense(units = 1))
```

10. Compile the network with an **adam** optimizer and use **Mean Squared Error** for the loss. Fit the model to the training data for **100** epochs with a batch size of **32**:

```
# Compiling the RNN
model.compile(optimizer = 'adam', loss = 'mean_squared_error')

# Fitting the RNN to the Training set
model.fit(X_train, y_train, epochs = 100, batch_size = 32)
```

11. Load and process the test data (which is treated as actual data here) and select the column representing the value of **Open** stock data:

```
dataset_testing = pd.read_csv("../GOOG_test.csv")
actual_stock_price = dataset_testing[['Open']].values
actual_stock_price
```

12. Concatenate the data since we will need **60** previous instances to get the stock price for each day. Therefore, we will need both the training and test data:

```
total_data = pd.concat((dataset_training['Open'], \
                        dataset_testing['Open']), axis = 0)
```

13. Reshape and scale the input to prepare the test data. Note that we are predicting the January monthly trend, which has **21** financial days, so in order to prepare the test set, we take the lower bound value as **60** and the upper bound value as **81**. This ensures that the difference of **21** is maintained:

```
inputs = total_data[len(total_data) \
                    - len(dataset_testing) - 60:].values
inputs = inputs.reshape(-1,1)
inputs = sc.transform(inputs)
X_test = []
for i in range(60, 81):
    X_test.append(inputs[i-60:i, 0])
X_test = np.array(X_test)
X_test = np.reshape(X_test, (X_test.shape[0], \
                    X_test.shape[1], 1))
predicted_stock_price = model.predict(X_test)
predicted_stock_price = sc.inverse_transform(\
                        predicted_stock_price)
```

14. Visualize the results by plotting the actual stock price and plotting the predicted stock price:

```
# Visualizing the results
plt.plot(actual_stock_price, color = 'green', \
         label = 'Real Alphabet Stock Price',ls='--')
plt.plot(predicted_stock_price, color = 'red', \
         label = 'Predicted Alphabet Stock Price',ls='-')
plt.title('Predicted Stock Price')
plt.xlabel('Time in days')
plt.ylabel('Real Stock Price')
plt.legend()
plt.show()
```

Expected output:

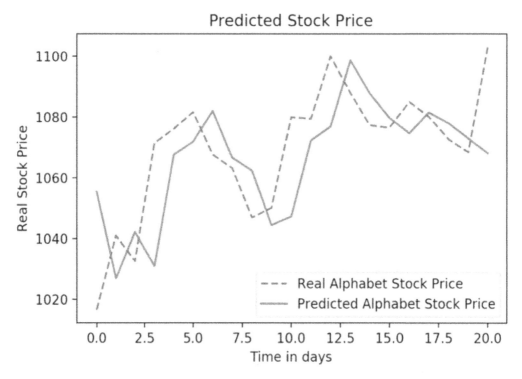

Figure 9.20: Real versus predicted stock price

> **NOTE**
>
> To access the source code for this specific section, please refer to
> https://packt.live/2ZDggf4.
>
> You can also run this example online at https://packt.live/2O4ZoJ7.

Now, if we compare the **LSTM** of *Exercise 9.01, Predicting the Trend of Alphabet's Stock Price Using an LSTM with 50 Units (Neurons)*, which had **50** neurons (units), with this **LSTM**, which uses **100** units, we can see that, unlike in the case of the Amazon stock price, the Alphabet stock trend is captured better using an **LSTM** with **100** units:

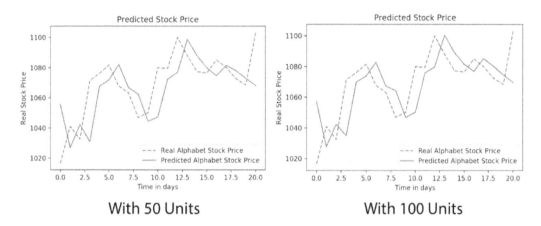

With 50 Units   With 100 Units

Figure 9.21: Comparing the output with the LSTM of Exercise 9.01

Thus, we can clearly see that an **LSTM** with **100** units predicts a more accurate trend than an **LSTM** with **50** units. Do keep in mind that an **LSTM** with **100** units will need more computational time but provides better results in this scenario. As well as modifying our model by adding more units, we can also add regularization. The following activity will test whether adding regularization can make our Amazon model more accurate.

ACTIVITY 9.02: PREDICTING AMAZON'S STOCK PRICE WITH ADDED REGULARIZATION

In this activity, we will examine the stock price of Amazon over the last 5 years, from January 1, 2014, to December 31, 2018. In doing so, we will try to predict and forecast the company's future trend for January 2019 using RNNs and an LSTM. We have the actual values for January 2019, so we will be able to compare our predictions with the actual values later. Initially, we predicted the trend of Amazon's stock price using an LSTM with 50 units (or neurons). Here, we will also add dropout regularization and compare the results with *Activity 9.01, Predicting the Trend of Amazon's Stock Price Using an LSTM with 50 Units (Neurons)*. Follow these steps to complete this activity:

1. Import the required libraries.

2. From the full dataset, extract the **Open** column since the predictions will be made on the open stock value. You can download the dataset from this book's GitHub repository at https://packt.live/2vtdA8o.

3. Normalize the data between 0 and 1.

4. Then, create timestamps. The values of each day in January 2019 will be predicted by the previous **60** days. So, if January 1 is predicted by using the value from the nth day up to December 31, then January 2 will be predicted by using the $n + 1^{st}$ day and January 1, and so on.

5. Reshape the data into three dimensions since the network needs the data to be in three dimensions.

6. Build an RNN with four LSTM layers in Keras, each with **50** units (here, units refer to neurons), and a 20% dropout after each LSTM layer. The first step should provide the input shape. Note that the final LSTM layer always adds `return_sequences=True`.

7. Process and prepare the test data, which is the actual data for January 2019.

8. Combine and process the train and test data.

9. Finally, visualize the results.

After implementing these steps, you should get the following expected output:

Figure 9.22: Real versus predicted stock prices

> **NOTE**
>
> The solution for this activity can be found on page 457.

In the next activity, we will experiment with building an **RNN** with **100** units in each **LSTM** layer and compare this with how the **RNN** performed with only **50** units.

ACTIVITY 9.03: PREDICTING THE TREND OF AMAZON'S STOCK PRICE USING AN LSTM WITH AN INCREASING NUMBER OF LSTM NEURONS (100 UNITS)

In this activity, we will examine the stock price of Amazon over the last 5 years, from January 1, 2014, to December 31, 2018. In doing so, we will try to predict and forecast the company's future trend for January 2019 using RNNs. We have the actual values for January 2019, so we will be able to compare our predictions with the actual values later. You can also compare the output difference with *Activity 9.01, Predicting the Trend of Amazon's Stock Price Using an LSTM with 50 Units (Neurons)*. Follow these steps to complete this activity:

1. Import the required libraries.

2. From the full dataset, extract the **Open** column since the predictions will be made on the **Open** stock value.

3. Normalize the data between 0 and 1.

4. Then, create timestamps. The values of each day in January 2019 will be predicted by the previous **60** days; so, if January 1 is predicted by using the value from the nth day up to December 31, then January 2 will be predicted by using the $n + 1^{st}$ day and January 1, and so on.

5. Reshape the data into three dimensions since the network needs data to be in three dimensions.

6. Build an LSTM in Keras with 100 units (here, units refer to neurons). The first step should provide the input shape. Note that the final **LSTM** layer always adds `return_sequences=True`. Compile and fit the model to the training data.

7. Process and prepare the test data, which is the actual data for January 2019.

8. Combine and process the training and test data.

9. Visualize the results.

After implementing these steps, you should get the following expected output:

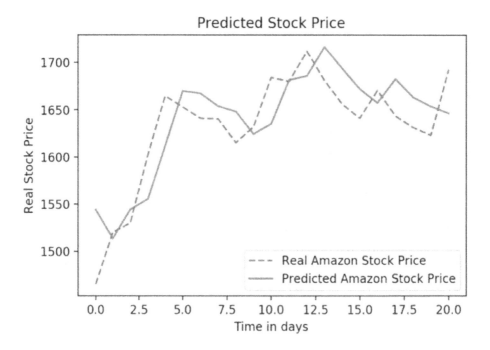

Figure 9.23: Real versus predicted stock prices

> **NOTE**
>
> The solution for this activity can be found on page 462.

In this activity, we created an **RNN** with four **LSTM** layers, each with **100** units. We compared this to the results of *Activity 9.02, Predicting Amazon's Stock Price with Added Regularization*, in which there were **50** units per layer. The difference between the two models was minimal, so a model with fewer units is preferable due to the decrease in computational time and there being a smaller possibility of overfitting the training data.

SUMMARY

In this chapter, we learned about sequential modeling and sequential memory by examining some real-life cases with Google Assistant. Then, we learned how sequential modeling is related to **RNNs**, as well as how **RNNs** are different from traditional feedforward networks. We learned about the vanishing gradient problem in detail and how using an **LSTM** is better than a simple **RNN** to overcome the vanishing gradient problem. We applied what we learned to time series problems by predicting stock trends.

In this workshop, we learned the basics of machine learning and Python, while also gaining an in-depth understanding of applying Keras to develop efficient deep learning solutions. We explored the difference between machine and deep learning. We began the workshop by building a logistic regression model, first with scikit-learn, and then with Keras.

Then, we explored Keras and its different models further by creating prediction models for various real-world scenarios, such as classifying online shoppers into those with purchase intention and those without. We learned how to evaluate, optimize, and improve models to achieve maximum information to create robust models that perform well on new, unseen data.

We also incorporated cross-validation by building Keras models with wrappers for scikit-learn that help those familiar with scikit-learn workstreams utilize Keras models easily. Then, we learned how to apply **L1**, **L2**, and `dropout regularization` techniques to improve the accuracy of models and to help prevent our models from overfitting the training data.

Next, we explored model evaluation further by applying techniques such as null accuracy for baseline comparison and evaluation metrics such as precision, the **AUC-ROC** score, and more to understand how our model scores classification tasks. Ultimately, these advanced evaluation techniques helped us understand under what conditions our model is performing well and where there is room for improvement.

We ended the workshop by creating some advanced models with Keras. We explored computer vision by building **CNN** models with various parameters to classify images. Then, we used pre-trained models to classify new images and fine-tuned those pre-trained models so that we could utilize them for our own applications. Finally, we covered sequential modeling, which is used for modeling sequences such as stock prices and natural language processing. We tested this knowledge by creating **RNN** networks with **LSTM** layers to predict the stock price of real stock data and experimented with various numbers of units in each layer and the effect of dropout regularization on the model's performance.

Overall, we have gained a comprehensive understanding of how to use Keras to solve a variety of problems using real-world datasets. We covered the classification tasks of online shoppers, hepatitis C data, and failure data for Scania trucks, as well as regression tasks such as predicting the aquatic toxicity of various chemicals when given various chemical attributes. We also performed image classification tasks and built **CNN** models to predict whether images are of flowers or cars, and also built regression tasks to predict future stock prices with **RNNs**. By using this workshop to build models with real-word datasets, you are ready to apply your learning and understanding to your own problem-solving and create your own applications.

APPENDIX

CHAPTER 1: INTRODUCTION TO MACHINE LEARNING WITH KERAS

ACTIVITY 1.01: ADDING REGULARIZATION TO THE MODEL

In this activity, we will utilize the same logistic regression model from the scikit-learn package. This time, however, we will add regularization to the model and search for the optimum regularization parameter - a process often called **hyperparameter tuning**. After training the models, we will test the predictions and compare the model evaluation metrics to the ones that were produced by the baseline model and the model without regularization.

1. Load the feature data from *Exercise 1.03*, *Appropriate Representation of the Data*, and the target data from *Exercise 1.02*, *Cleaning the Data*:

```
import pandas as pd
feats = pd.read_csv('../data/OSI_feats_e3.csv')
target = pd.read_csv('../data/OSI_target_e2.csv')
```

2. Create a **test** and **train** dataset. Train the data using the training dataset. This time, however, use part of the **training** dataset for validation in order to choose the most appropriate hyperparameter.

   Once again, we will use **test_size = 0.2**, which means that **20%** of the data will be reserved for testing. The size of our validation set will be determined by how many validation folds we have. If we do **10-fold cross-validation**, this equates to reserving **10%** of the **training** dataset to validate our model on. Each fold will use a different **10%** of the **training** dataset, and the average error across all folds is used to compare models with different hyperparameters. Assign a random value to the **random_state** variable:

```
from sklearn.model_selection import train_test_split
test_size = 0.2
random_state = 13
X_train, X_test, y_train, y_test = \
train_test_split(feats, target, test_size=test_size, \
                 random_state=random_state)
```

3. Check the dimensions of the DataFrames:

```
print(f'Shape of X_train: {X_train.shape}')
print(f'Shape of y_train: {y_train.shape}')
print(f'Shape of X_test: {X_test.shape}')
print(f'Shape of y_test: {y_test.shape}')
```

The preceding code produces the following output:

```
Shape of X_train: (9864, 68)
Shape of y_train: (9864, 1)
Shape of X_test: (2466, 68)
Shape of y_test: (2466, 1)
```

4. Next, instantiate the models. Try two types of regularization parameters, **l1** and **l2**, with 10-fold cross-validation. Iterate our regularization parameter from 1×10^{-2} to 1×10^6 equally in the logarithmic space to observe how the parameters affect the results:

```
import numpy as np
from sklearn.linear_model import LogisticRegressionCV
Cs = np.logspace(-2, 6, 9)
model_l1 = LogisticRegressionCV(Cs=Cs, penalty='l1', \
                                cv=10, solver='liblinear', \
                                random_state=42, max_iter=10000)
model_l2 = LogisticRegressionCV(Cs=Cs, penalty='l2', cv=10, \
                                random_state=42, max_iter=10000)
```

> **NOTE**
>
> For a logistic regression model with the **l1** regularization parameter, only the **liblinear** solver can be used.

5. Next, fit the models to the training data:

```
model_l1.fit(X_train, y_train['Revenue'])
model_l2.fit(X_train, y_train['Revenue'])
```

The following figure shows the output of the preceding code:

```
LogisticRegressionCV(Cs=array([1.e-02, 1.e-01, 1.e+00, 1.e+01, 1.e+02, 1.e+03, 1.e+04, 1.e+05,
       1.e+06]),
                     class_weight=None, cv=10, dual=False, fit_intercept=True,
                     intercept_scaling=1.0, l1_ratios=None, max_iter=10000,
                     multi_class='auto', n_jobs=None, penalty='l2',
                     random_state=42, refit=True, scoring=None, solver='lbfgs',
                     tol=0.0001, verbose=0)
```

Figure 1.37: Output of the fit command indicating all of the model training parameters

6. Here, we can see what the value of the regularization parameter was for the two different models. The regularization parameter is chosen according to which produced a model with the lowest error:

```
print(f'Best hyperparameter for l1 regularization model: \
{model_l1.C_[0]}')
print(f'Best hyperparameter for l2 regularization model: \
{model_l2.C_[0]}')
```

The preceding code produces the following output:

```
Best hyperparameter for l1 regularization model: 1000000.0
Best hyperparameter for l2 regularization model: 1.0
```

> **NOTE**
>
> The C_ attribute is only available once the model has been trained because it is set once the best parameter from the cross-validation process has been determined.

7. To evaluate the performance of the models, make predictions on the **test** set, which we'll compare against the **true** values:

```
y_pred_l1 = model_l1.predict(X_test)
y_pred_l2 = model_l2.predict(X_test)
```

8. To compare these models, calculate the evaluation metrics. First, look at the accuracy of the model:

```
from sklearn import metrics
accuracy_l1 = metrics.accuracy_score(y_pred=y_pred_l1, \
                                     y_true=y_test)
accuracy_l2 = metrics.accuracy_score(y_pred=y_pred_l2, \
                                     y_true=y_test)
print(f'Accuracy of the model with l1 regularization is \
{accuracy_l1*100:.4f}%')
print(f'Accuracy of the model with l2 regularization is \
{accuracy_l2*100:.4f}%')
```

The preceding code produces the following output:

```
Accuracy of the model with l1 regularization is 89.2133%
Accuracy of the model with l2 regularization is 89.2944%
```

9. Also, look at the other evaluation metrics:

```
precision_l1, recall_l1, fscore_l1, _ = \
metrics.precision_recall_fscore_support(y_pred=y_pred_l1, \
                                        y_true=y_test, \
                                        average='binary')

precision_l2, recall_l2, fscore_l2, _ = \
metrics.precision_recall_fscore_support(y_pred=y_pred_l2, \
                                        y_true=y_test, \
                                        average='binary')

print(f'l1\nPrecision: {precision_l1:.4f}\nRecall: \
{recall_l1:.4f}\nfscore: {fscore_l1:.4f}\n\n')
print(f'l2\nPrecision: {precision_l2:.4f}\nRecall: \
{recall_l2:.4f}\nfscore: {fscore_l2:.4f}')
```

The preceding code produces the following output:

```
l1
Precision: 0.7300
Recall: 0.4078
fscore: 0.5233

l2
Precision: 0.7350
Recall: 0.4106
fscore: 0.5269
```

10. Observe the values of the coefficients once the model has been trained:

```
coef_list = [f'{feature}: {coef}' for coef, \
             feature in sorted(zip(model_l1.coef_[0], \
                               X_train.columns.values.tolist()))]
for item in coef_list:
    print(item)
```

> **NOTE**
>
> The **coef_** attribute is only available once the model has been trained because it is set once the best parameter from the cross-validation process has been determined.

The following figure shows the output of the preceding code:

```
ExitRates: -15.883224778812533
TrafficType_15: -14.795568293882813
TrafficType_19: -14.572417554129022
Browser_11: -14.194513734001333
TrafficType_18: -13.580838983781868
TrafficType_12: -10.887010740988496
Browser_3: -1.591759861518971
OperatingSystems_6: -1.3061883214341672
Browser_13: -1.1580420838412984
TrafficType_13: -1.1370525347807676
TrafficType_14: -1.0975546272442658
Browser_6: -1.0010333108800196
BounceRates: -0.9141131252024657
Browser_7: -0.8502023615800486
TrafficType_3: -0.811958144680347
TrafficType_6: -0.610088799226249
OperatingSystems_3: -0.608450446552214
TrafficType_1: -0.6050916722027272
OperatingSystems_1: -0.5365902656959869
OperatingSystems_4: -0.5111451125487375
TrafficType_4: -0.5062856279182754
Browser_2: -0.5059966973246434
TrafficType_2: -0.46430066907789846
TrafficType_9: -0.4466955982654509
Browser_4: -0.4214541023779841
Browser_5: -0.41616336572267953
TrafficType_5: -0.415441717179765
TrafficType_7: -0.3955699524561608
Browser_1: -0.3880867021005534
Browser_8: -0.3829361496965676
OperatingSystems_2: -0.3452133463717456
TrafficType_10: -0.26699389912184146
TrafficType_11: -0.20084569889820056
SpecialDay: -0.16993604135246643
VisitorType_Returning_Visitor: -0.1585189488537066
Region_9: -0.1357553937048372
Browser_10: -0.11275136307019001
Region_4: -0.07146513811457653
TrafficType_20: -0.027994440964808312
TrafficType_8: -0.01080321579230938
Administrative: -0.008848764565436021
ProductRelated_Duration: 7.884230862995893e-05
Administrative_Duration: 0.00010234279673558182
Informational_Duration: 0.0003083388371428332
ProductRelated: 0.0011034989322559698
Informational: 0.00462053503124074
OperatingSystems_8: 0.014922953675113934
VisitorType_New_Visitor: 0.015142798459659391
Region_7: 0.035576200252320694
Region_1: 0.03654049619361953
Region_3: 0.06097956829573871
PageValues: 0.08295642466041496
is_weekend: 0.12026927900778797
Region_8: 0.15007889661364254
Region_6: 0.1946605704463812
Region_2: 0.2409191030186223
OperatingSystems_7: 0.6430414471829574
Month_Dec: 0.9259040930136528
Month_Mar: 0.9329536574050367
Month_May: 1.0175770611658324
TrafficType_16: 1.219454600596873
Month_June: 1.3396248804059154
Month_Aug: 1.383936112795327
Month_Oct: 1.4462669163574475
Browser_12: 1.521383337562806
Month_Sep: 1.527112925006471
Month_Jul: 1.7056603829598316
Month_Nov: 2.0227013012240787
```

Figure 1.38: The feature column names and the value of their respective coefficients for the model with l1 regularization

11. Do the same for the model with an **12** regularization parameter type:

```
coef_list = [f'{feature}: {coef}' for coef, \
             feature in sorted(zip(model_l2.coef_[0], \
                               X_train.columns.values.tolist()))]
for item in coef_list:
    print(item)
```

The following figure shows the output of the preceding code:

```
TrafficType_13: -0.34009141528704256
Month_May: -0.2996216354738292
TrafficType_3: -0.26886567044331466
Month_Dec: -0.2600371384272894
Month_Mar: -0.234528066357572
VisitorType_Returning_Visitor: -0.2092729117037135
ExitRates: -0.20125071612863224
OperatingSystems_3: -0.17171727353008637
BounceRates: -0.15860752704242273
SpecialDay: -0.15806189654702138
TrafficType_1: -0.14304912321990237
Browser_6: -0.10194914517637654
Region_4: -0.09904631488404457
Region_9: -0.09873197637696395
Browser_3: -0.08914198358022751
Month_June: -0.042681926326746056
TrafficType_15: -0.04024689666293222
Browser_7: -0.03327911496921953
TrafficType_6: -0.031102643479272572
Region_1: -0.027052218462517107
OperatingSystems_6: -0.02574568309220721
TrafficType_19: -0.02203436126130139
Browser_13: -0.021797739222369092
Browser_2: -0.017512124173199008
Region_7: -0.010600896457621618
TrafficType_14: -0.009414755781185648
TrafficType_18: -0.008565351627323418
Browser_11: -0.007825344595778813
Region_3: -0.00722706571136333
Informational: -0.002843129468954964
OperatingSystems_1: -0.0027540039503951706
TrafficType_9: -0.00239566032775648
OperatingSystems_8: -0.0016314917865704064
TrafficType_12: -0.0010615530198468937
OperatingSystems_4: 1.6377699949524623e-05
ProductRelated_Duration: 6.203896052153687e-05
Administrative_Duration: 0.0001498531976570715
Informational_Duration: 0.0002646491833362014
ProductRelated: 0.003308445071856749
Browser_8: 0.0054882974452588875
Month_Aug: 0.009417525299432315
TrafficType_7: 0.009865390498040204
OperatingSystems_7: 0.0106858800764906
Administrative: 0.011818428072388093
Browser_5: 0.011917760976019723
TrafficType_16: 0.016810452784330476
Browser_1: 0.022909998435860734
Region_8: 0.023998792506279735
Browser_4: 0.03535665081078685
Browser_12: 0.042525537053214074
TrafficType_5: 0.050267199536111515
TrafficType_4: 0.06057990490809306
Month_Oct: 0.07064830444129157
Region_6: 0.0736506138448771
Browser_10: 0.0810882098129662
TrafficType_20: 0.08608941436142269
Month_Sep: 0.0878149242855243
PageValues: 0.08801299639787297
TrafficType_11: 0.0991850902887261
Region_2: 0.11817125757973408
TrafficType_2: 0.1195593468057169
OperatingSystems_2: 0.12562135700649213
TrafficType_10: 0.12594481607945368
Month_Jul: 0.13420860404449297
is_weekend: 0.15588521706821454
VisitorType_New_Visitor: 0.18484311217638447
TrafficType_8: 0.22599108047373068
Month_Nov: 0.6282700212046415
```

Figure 1.39: The feature column names and the value of their respective coefficients for the model with l2 regularization

NOTE

To access the source code for this specific section, please refer to https://packt.live/2Vloe5M.

This section does not currently have an online interactive example, and will need to be run locally.

CHAPTER 2: MACHINE LEARNING VERSUS DEEP LEARNING

ACTIVITY 2.01: CREATING A LOGISTIC REGRESSION MODEL USING KERAS

In this activity, we are going to create a basic model using the Keras library. The model that we will build will classify users of a website into those that will purchase a product from a website and those that will not. To do this, we will utilize the same online shopping purchasing intention dataset that we did previously and attempt to predict the same variables that we did in *Chapter 1, Introduction to Machine Learning with Keras.*

Perform the following steps to complete this activity:

1. Open a Jupyter notebook from the start menu to implement this activity. Load in the online shopping purchasing intention datasets, which you can download from the GitHub repository. We will use the pandas library for data loading, so import the **pandas** library. Ensure you have saved the csv files to an appropriate data folder for this chapter first. Alternatively, you can change the path to the files that you use in your code.

```
import pandas as pd
feats = pd.read_csv('../data/OSI_feats.csv')
target = pd.read_csv('../data/OSI_target.csv')
```

2. For the purposes of this activity, we will not perform any further preprocessing. As we did in the previous chapter, we will split the dataset into training and testing and leave the testing until the very end when we evaluate our models. We will reserve **20%** of our data for testing by setting the **test_size=0.2** parameter, and we will create a **random_state** parameter so that we can recreate the results:

```
from sklearn.model_selection import train_test_split
test_size = 0.2
random_state = 42

X_train, X_test, y_train, y_test = \
train_test_split(feats, target, test_size=test_size, \
                 random_state=random_state)
```

3. Set a seed in **numpy** and **tensorflow** for reproducibility. Begin creating the model by initializing a model of the **Sequential** class:

```
from keras.models import Sequential
import numpy as np
from tensorflow import random
np.random.seed(random_state)
random.set_seed(random_state)
model = Sequential()
```

4. To add a fully connected layer to the model, add a layer of the **Dense** class. Here, we include the number of nodes in the layer. In our case, this will be one since we are performing binary classification and our desired output is **zero** or **one**. Also, specify the input dimensions, which is only done on the first layer of the model. It is there to indicate the format of the input data. Pass the number of features:

```
from keras.layers import Dense
model.add(Dense(1, input_dim=X_train.shape[1]))
```

5. Add a sigmoid activation function to the output of the previous layer to replicate the **logistic regression** algorithm:

```
from keras.layers import Activation
model.add(Activation('sigmoid'))
```

6. Once we have all the model components in the correct order, we must compile the model so that all the learning processes are configured. Use the **adam** optimizer, a **binary_crossentropy** for the loss, and track the accuracy of the model by passing the parameter into the **metrics** argument:

```
model.compile(optimizer='adam', loss='binary_crossentropy', \
              metrics=['accuracy'])
```

7. Print the model summary to verify the model is as we expect it to be:

```
print(model.summary())
```

The following figure shows the output of the preceding code:

```
Model: "sequential_1"
```

Layer (type)	Output Shape	Param #
dense_1 (Dense)	(None, 1)	69
activation_1 (Activation)	(None, 1)	0

```
Total params: 69
Trainable params: 69
Non-trainable params: 0
```

```
None
```

Figure 2.19: A summary of the model

8. Next, fit the model using the **fit** method of the **model** class. Provide the training data, as well as the number of epochs and how much data to use for validation after each epoch:

```
history = model.fit(X_train, y_train['Revenue'], epochs=10, \
                    validation_split=0.2, shuffle=False)
```

The following figure shows the output of the preceding code:

```
Train on 7891 samples, validate on 1973 samples
Epoch 1/10
7891/7891 [==============================] - 0s 49us/step - loss: 3.4358 - accuracy: 0.7656 - val_loss: 0.9237 - val_
accuracy: 0.8702
Epoch 2/10
7891/7891 [==============================] - 0s 34us/step - loss: 0.8518 - accuracy: 0.8446 - val_loss: 0.7980 - val_
accuracy: 0.8920
Epoch 3/10
7891/7891 [==============================] - 0s 36us/step - loss: 0.5456 - accuracy: 0.8680 - val_loss: 0.5081 - val_
accuracy: 0.8890
Epoch 4/10
7891/7891 [==============================] - 0s 37us/step - loss: 0.4471 - accuracy: 0.8761 - val_loss: 0.3054 - val_
accuracy: 0.8946
Epoch 5/10
7891/7891 [==============================] - 0s 34us/step - loss: 0.3870 - accuracy: 0.8808 - val_loss: 0.3196 - val_
accuracy: 0.8844
Epoch 6/10
7891/7891 [==============================] - 0s 40us/step - loss: 0.3938 - accuracy: 0.8764 - val_loss: 0.3646 - val_
accuracy: 0.8819
Epoch 7/10
7891/7891 [==============================] - 0s 35us/step - loss: 0.3527 - accuracy: 0.8832 - val_loss: 0.3138 - val_
accuracy: 0.8855
Epoch 8/10
7891/7891 [==============================] - 0s 36us/step - loss: 0.3350 - accuracy: 0.8813 - val_loss: 0.3891 - val_
accuracy: 0.8789
Epoch 9/10
7891/7891 [==============================] - 0s 38us/step - loss: 0.3477 - accuracy: 0.8799 - val_loss: 0.3302 - val_
accuracy: 0.8814
Epoch 10/10
7891/7891 [==============================] - 0s 34us/step - loss: 0.3640 - accuracy: 0.8816 - val_loss: 0.3136 - val_
accuracy: 0.8839
```

Figure 2.20: Using the fit method on the model

9. The values for the loss and accuracy have been stored within the **history** variable. Plot the values for each using the loss and accuracy we tracked after each epoch:

```
import matplotlib.pyplot as plt
%matplotlib inline

# Plot training and validation accuracy values
plt.plot(history.history['accuracy'])
plt.plot(history.history['val_accuracy'])
plt.title('Model accuracy')
plt.ylabel('Accuracy')
plt.xlabel('Epoch')
plt.legend(['Train', 'Validation'], loc='upper left')
plt.show()

# Plot training and validation loss values
plt.plot(history.history['loss'])
plt.plot(history.history['val_loss'])
plt.title('Model loss')
plt.ylabel('Loss')
plt.xlabel('Epoch')
plt.legend(['Train', 'Validation'], loc='upper left')
plt.show()
```

The following plots show the output of the preceding code:

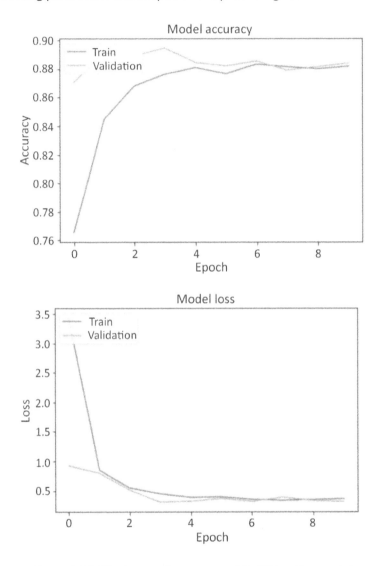

Figure 2.21: The loss and accuracy while fitting the model

10. Finally, evaluate the model on the test data we held out from the beginning, which will give an objective evaluation of the performance of the model:

```
test_loss, test_acc = model.evaluate(X_test, y_test['Revenue'])
print(f'The loss on the test set is {test_loss:.4f} \
and the accuracy is {test_acc*100:.3f}%')
```

The output of the preceding code can be found below. Here, the model predicts the purchasing intention of users in the test dataset and evaluates the performance by comparing it to the real values in **y_test**. Evaluating the model on the test dataset produces loss and accuracy values that we can print out:

```
2466/2466 [==============================] - 0s 15us/step
The loss on the test set is 0.3632 and the accuracy is 86.902%
```

> **NOTE**
>
> To access the source code for this specific section, please refer to
> https://packt.live/3dVTQLe.
>
> You can also run this example online at https://packt.live/2ZxEhV4.

CHAPTER 3: DEEP LEARNING WITH KERAS

ACTIVITY 3.01: BUILDING A SINGLE-LAYER NEURAL NETWORK FOR PERFORMING BINARY CLASSIFICATION

In this activity, we will compare the results of a logistic regression model and single-layer neural networks of different node sizes and different activation functions. The dataset we will use represents the normalized test results of aircraft propeller inspections, while the class represents whether they passed or failed a manual visual inspection. We will create models to predict the results of the manual inspection when given the automated test results. Follow these steps to complete this activity:

1. Load all the required packages:

```
# import required packages from Keras
from keras.models import Sequential
from keras.layers import Dense, Activation
import numpy as np
import pandas as pd
from tensorflow import random
from sklearn.model_selection import train_test_split
# import required packages for plotting
import matplotlib.pyplot as plt
import matplotlib
%matplotlib inline
import matplotlib.patches as mpatches
# import the function for plotting decision boundary
from utils import plot_decision_boundary
```

2. Set up a **seed**:

```
"""
define a seed for random number generator so the result will be
reproducible
"""
seed = 1
```

3. Load the simulated dataset and print the size of **X** and **Y** and the number of examples:

```
"""
load the dataset, print the shapes of input and output and the number
of examples
"""
feats = pd.read_csv('../data/outlier_feats.csv')
target = pd.read_csv('../data/outlier_target.csv')
print("X size = ", feats.shape)
print("Y size = ", target.shape)
print("Number of examples = ", feats.shape[0])
```

Expected output:

```
X size = (3359, 2)
Y size = (3359, 1)
Number of examples = 3359
```

4. Plot the dataset. The x and y coordinates of each point will be the two input features. The color of each record represents the **pass/fail** result:

```
class_1=plt.scatter(feats.loc[target['Class']==0,'feature1'], \
                    feats.loc[target['Class']==0,'feature2'], \
                    c="red", s=40, edgecolor='k')

class_2=plt.scatter(feats.loc[target['Class']==1,'feature1'], \
                    feats.loc[target['Class']==1,'feature2'], \
                    c="blue", s=40, edgecolor='k')

plt.legend((class_1, class_2),('Fail','Pass'))
plt.xlabel('Feature 1')
plt.ylabel('Feature 2')
```

The following image shows the output of the preceding code:

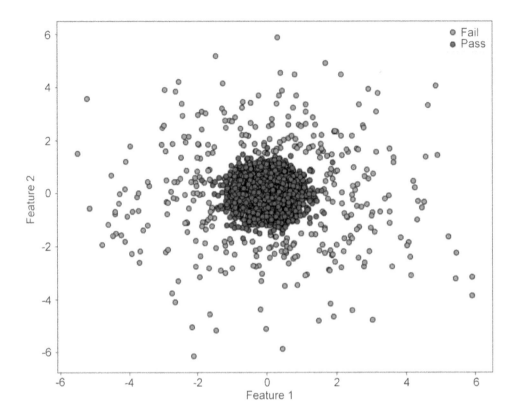

Figure 3.19: Simulated training data points

5. Build the **logistic regression** model, which will be a one-node sequential model with no hidden layers and a **sigmoid activation** function:

```
np.random.seed(seed)
random.set_seed(seed)
model = Sequential()
model.add(Dense(1, activation='sigmoid', input_dim=2))
model.compile(optimizer='sgd', loss='binary_crossentropy')
```

6. Fit the model to the training data:

```
model.fit(feats, target, batch_size=5, epochs=100, verbose=1, \
          validation_split=0.2, shuffle=False)
```

Expected output:

The loss on the validation set after **100** epochs = **0.3537**:

```
Epoch 96/100
2687/2687 [==============================] - 0s 159us/step - loss: 0.3365 - val_loss: 0.3546
Epoch 97/100
2687/2687 [==============================] - 0s 153us/step - loss: 0.3366 - val_loss: 0.3545
Epoch 98/100
2687/2687 [==============================] - 0s 156us/step - loss: 0.3365 - val_loss: 0.3541
Epoch 99/100
2687/2687 [==============================] - 0s 155us/step - loss: 0.3365 - val_loss: 0.3536
Epoch 100/100
2687/2687 [==============================] - 0s 156us/step - loss: 0.3366 - val_loss: 0.3537
```

Figure 3.20: The loss details of the last 5 epochs out of 100

7. Plot the decision boundary on the training data:

```
matplotlib.rcParams['figure.figsize'] = (10.0, 8.0)
plot_decision_boundary(lambda x: model.predict(x), feats, target)
plt.title("Logistic Regression")
```

The following image shows the output of the preceding code:

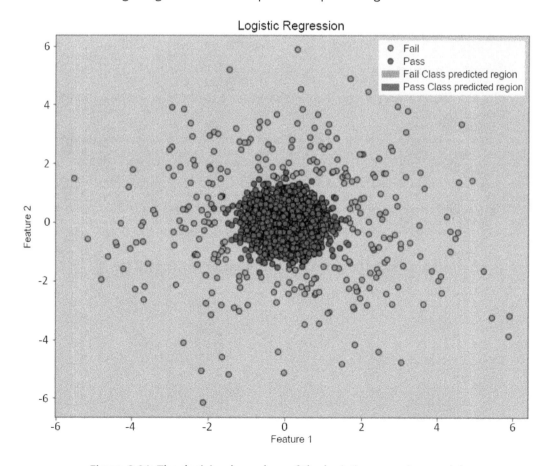

Figure 3.21: The decision boundary of the logistic regression model

The linear decision boundary of the logistic regression model is obviously unable to capture the circular decision boundary between the two classes and predicts all the results as a passed result.

8. Create a neural network with one hidden layer with three nodes and a **relu activation function** and an output layer with one node and a **sigmoid activation function**. Finally, compile the model:

```
np.random.seed(seed)
random.set_seed(seed)
model = Sequential()
model.add(Dense(3, activation='relu', input_dim=2))
model.add(Dense(1, activation='sigmoid'))
model.compile(optimizer='sgd', loss='binary_crossentropy')
```

9. Fit the model to the training data:

```
model.fit(feats, target, batch_size=5, epochs=200, verbose=1, \
          validation_split=0.2, shuffle=False)
```

Expected output:

The loss that's evaluated on the validation set after **200** epochs = **0.0260**:

```
Epoch 196/200
2687/2687 [==============================] - 0s 163us/step - loss: 0.0131 - val_loss: 0.0261
Epoch 197/200
2687/2687 [==============================] - 0s 163us/step - loss: 0.0130 - val_loss: 0.0261
Epoch 198/200
2687/2687 [==============================] - 0s 165us/step - loss: 0.0130 - val_loss: 0.0259
Epoch 199/200
2687/2687 [==============================] - 0s 169us/step - loss: 0.0130 - val_loss: 0.0259
Epoch 200/200
2687/2687 [==============================] - 0s 161us/step - loss: 0.0129 - val_loss: 0.0260
```

Figure 3.22: The loss details of the last 5 epochs out of 200

10. Plot the decision boundary that was created:

```
matplotlib.rcParams['figure.figsize'] = (10.0, 8.0)
plot_decision_boundary(lambda x: model.predict(x), feats, target)
plt.title("Decision Boundary for Neural Network with "\
        "hidden layer size 3")
```

The following image shows the output of the preceding code:

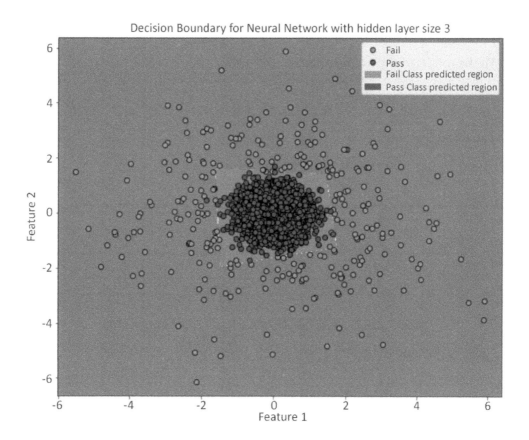

Figure 3.23: The decision boundary for the neural network with a hidden layer size of 3 and a ReLU activation function

Having three processing units instead of one dramatically improved the capability of the model in capturing the non-linear boundary between the two classes. Notice that the loss value decreased drastically in comparison to the previous step.

11. Create a neural network with one hidden layer with six nodes and a **relu activation function** and an output layer with one node and a **sigmoid activation function**. Finally, compile the model:

```
np.random.seed(seed)
random.set_seed(seed)
model = Sequential()
model.add(Dense(6, activation='relu', input_dim=2))
model.add(Dense(1, activation='sigmoid'))
model.compile(optimizer='sgd', loss='binary_crossentropy')
```

12. Fit the model to the training data:

```
model.fit(feats, target, batch_size=5, epochs=400, verbose=1, \
          validation_split=0.2, shuffle=False)
```

Expected output:

The loss after **400** epochs = **0.0231**:

```
Epoch 396/400
2687/2687 [==============================] - 0s 174us/step - loss: 0.0072 - val_loss: 0.0232
Epoch 397/400
2687/2687 [==============================] - 0s 166us/step - loss: 0.0072 - val_loss: 0.0233
Epoch 398/400
2687/2687 [==============================] - 0s 180us/step - loss: 0.0072 - val_loss: 0.0232
Epoch 399/400
2687/2687 [==============================] - 0s 164us/step - loss: 0.0072 - val_loss: 0.0232
Epoch 400/400
2687/2687 [==============================] - 0s 165us/step - loss: 0.0072 - val_loss: 0.0231
```

Figure 3.24: The loss details of the last 5 epochs out of 400

13. Plot the decision boundary:

```
matplotlib.rcParams['figure.figsize'] = (10.0, 8.0)
plot_decision_boundary(lambda x: model.predict(x), feats, target)
plt.title("Decision Boundary for Neural Network with "\
          "hidden layer size 6")
```

The following image shows the output of the preceding code:

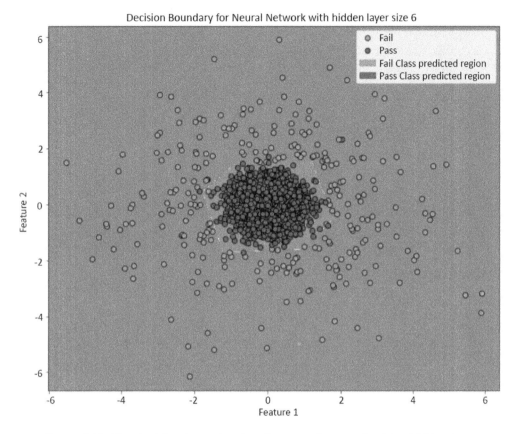

Figure 3.25: The decision boundary for the neural network with a hidden layer size of 6 and the ReLU activation function

By doubling the number of units in the hidden layer, the decision boundary of the model gets closer to a true circular shape, and the loss value is decreased even more in comparison to the previous step.

14. Create a neural network with one hidden layer with three nodes and a **tanh activation function** and an output layer with one node and a **sigmoid activation function**. Finally, compile the model:

```
np.random.seed(seed)
random.set_seed(seed)
model = Sequential()
model.add(Dense(3, activation='tanh', input_dim=2))
model.add(Dense(1, activation='sigmoid'))
model.compile(optimizer='sgd', loss='binary_crossentropy')
```

15. Fit the model to the training data:

```
model.fit(feats, target, batch_size=5, epochs=200, verbose=1, \
          validation_split=0.2, shuffle=False)
```

Expected output:

The loss after **200** epochs = **0.0426**:

```
Epoch 196/200
2687/2687 [==============================] - 0s 173us/step - loss: 0.0278 - val_loss: 0.0427
Epoch 197/200
2687/2687 [==============================] - 0s 160us/step - loss: 0.0277 - val_loss: 0.0426
Epoch 198/200
2687/2687 [==============================] - 0s 161us/step - loss: 0.0277 - val_loss: 0.0426
Epoch 199/200
2687/2687 [==============================] - 0s 166us/step - loss: 0.0276 - val_loss: 0.0426
Epoch 200/200
2687/2687 [==============================] - 0s 169us/step - loss: 0.0275 - val_loss: 0.0426
```

Figure 3.26: The loss details of the last 5 epochs out of 200

16. Plot the decision boundary:

```
plot_decision_boundary(lambda x: model.predict(x), feats, target)
plt.title("Decision Boundary for Neural Network with "\
          "hidden layer size 3")
```

The following image shows the output of the preceding code:

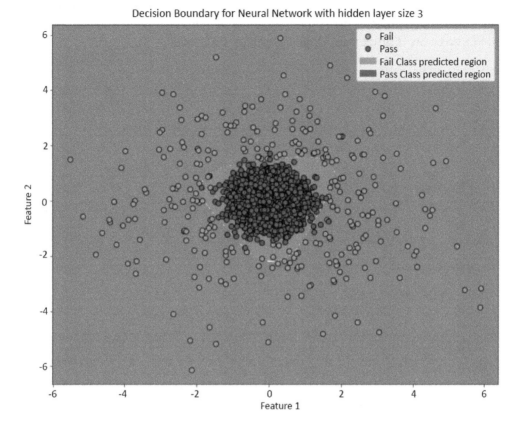

Figure 3.27: The decision boundary for the neural network with a hidden layer size of 3 and the tanh activation function

Using the **tanh** activation function has eliminated the sharp edges in the decision boundary. In other words, it has made the decision boundary smoother. However, the model is not performing better since we can see an increase in the loss value. We achieved similar loss and accuracy scores when we evaluated on the test dataset, despite mentioning previously that the learning parameters for **tanh** are slower than they are for **relu**.

17. Create a neural network with one hidden layer with six nodes and a **tanh activation function** and an output layer with one node and a **sigmoid activation function**. Finally, compile the model:

```
np.random.seed(seed)
random.set_seed(seed)
model = Sequential()
model.add(Dense(6, activation='tanh', input_dim=2))
model.add(Dense(1, activation='sigmoid'))
model.compile(optimizer='sgd', loss='binary_crossentropy')
```

18. Fit the model to the training data:

```
model.fit(feats, target, batch_size=5, epochs=400, verbose=1, \
          validation_split=0.2, shuffle=False)
```

Expected output:

The loss after **400** epochs = **0.0215**:

```
Epoch 396/400
2687/2687 [==============================] - 0s 168us/step - loss: 0.0140 - val_loss: 0.0216
Epoch 397/400
2687/2687 [==============================] - 0s 169us/step - loss: 0.0139 - val_loss: 0.0216
Epoch 398/400
2687/2687 [==============================] - 0s 169us/step - loss: 0.0139 - val_loss: 0.0216
Epoch 399/400
2687/2687 [==============================] - 0s 172us/step - loss: 0.0139 - val_loss: 0.0215
Epoch 400/400
2687/2687 [==============================] - 1s 209us/step - loss: 0.0139 - val_loss: 0.0215
```

Figure 3.28: The loss details of the last 5 epochs out of 400

19. Plot the decision boundary:

```
matplotlib.rcParams['figure.figsize'] = (10.0, 8.0)
plot_decision_boundary(lambda x: model.predict(x), feats, target)
plt.title("Decision Boundary for Neural Network with "\
          "hidden layer size 6")
```

The following image shows the output of the preceding code:

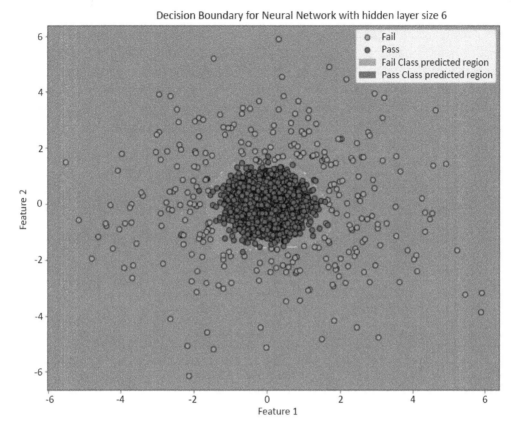

Figure 3.29: The decision boundary for the neural network with a hidden layer size of 6 and the tanh activation function

Again, using the **tanh** activation function instead of **relu** and adding more nodes to our hidden layer has smoothed the curves on the decision boundary more, fitting the training data better according to the accuracy of the training data. We should be careful not to add too many nodes to the hidden layer as we may begin to overfit the data. This can be observed by evaluating the test set, where there is a slight decrease in the accuracy of the neural network with six nodes compared to a neural network with three.

> **NOTE**
>
> To access the source code for this specific section, please refer to
> https://packt.live/3iv0wn1.
>
> You can also run this example online at https://packt.live/2BqumZt.

ACTIVITY 3.02: ADVANCED FIBROSIS DIAGNOSIS WITH NEURAL NETWORKS

In this activity, you are going to use a real dataset to predict whether a patient has advanced fibrosis based on measurements such as age, gender, and BMI. The dataset consists of information for 1,385 patients who underwent treatment dosages for hepatitis C. For each patient, **28** different attributes are available, as well as a class label, which can only take two values: **1**, indicating advanced fibrosis, and **0**, indicating no indication of advanced fibrosis. This is a binary/two-class classification problem with an input dimension equal to 28.

In this activity, you will implement different deep neural network architectures to perform this classification, plot the trends in training error rates and test error rates, and determine how many epochs the final classifier needs to be trained for. Follow these steps to complete this activity:

1. Import all the necessary libraries and load the dataset using the pandas **read_csv** function:

```
import pandas as pd
import numpy as np
from tensorflow import random
from sklearn.model_selection import train_test_split
from sklearn.preprocessing import StandardScaler
from keras.models import Sequential
from keras.layers import Dense

import matplotlib.pyplot as plt
import matplotlib
%matplotlib inline

X = pd.read_csv('../data/HCV_feats.csv')
y = pd.read_csv('../data/HCV_target.csv')
```

2. Print the number of **records** and **features** in the **feature** dataset and the number of unique classes in the **target** dataset:

```
print("Number of Examples in the Dataset = ", X.shape[0])
print("Number of Features for each example = ", X.shape[1])
print("Possible Output Classes = ", \
      y['AdvancedFibrosis'].unique())
```

Expected output:

```
Number of Examples in the Dataset = 1385
Number of Features for each example = 28
Possible Output Classes = [0 1]
```

3. Normalize the data and scale it. Following this, split the dataset into the **training** and **test** sets:

```
seed = 1
np.random.seed(seed)

sc = StandardScaler()
X = pd.DataFrame(sc.fit_transform(X), columns=X.columns)
X_train, X_test, y_train, y_test = \
train_test_split(X, y, test_size=0.2, random_state=seed)

# Print the information regarding dataset sizes
print(X_train.shape)
print(y_train.shape)
print(X_test.shape)
print(y_test.shape)
print ("Number of examples in training set = ", X_train.shape[0])
print ("Number of examples in test set = ", X_test.shape[0])
```

Expected output:

```
(1108, 28)
(1108, 1)
(277, 28)
(277, 1)
Number of examples in training set = 1108
Number of examples in test set = 277
```

4. Implement a deep neural network with one hidden layer of size **3** and a **tanh activation function**, an output layer with one node, and a **sigmoid activation function**. Finally, compile the model and print out a summary of the model:

```
np.random.seed(seed)
random.set_seed(seed)

# define the keras model
```

```
classifier = Sequential()
classifier.add(Dense(units = 3, activation = 'tanh', \
                    input_dim=X_train.shape[1]))

classifier.add(Dense(units = 1, activation = 'sigmoid'))
classifier.compile(optimizer = 'sgd', loss = 'binary_crossentropy', \
                    metrics = ['accuracy'])

classifier.summary()
```

The following image shows the output of the preceding code:

```
Model: "sequential_1"
```

Layer (type)	Output Shape	Param #
dense_1 (Dense)	(None, 3)	87
dense_2 (Dense)	(None, 1)	4

```
Total params: 91
Trainable params: 91
Non-trainable params: 0
```

Figure 3.30: The architecture of the neural network

5. Fit the model to the training data:

```
history=classifier.fit(X_train, y_train, batch_size = 20, \
                    epochs = 100, validation_split=0.1, \
                    shuffle=False)
```

6. Plot the **training error rate** and **test error rate** for every epoch:

```
plt.plot(history.history['loss'])
plt.plot(history.history['val_loss'])
plt.ylabel('loss')
plt.xlabel('epoch')
plt.legend(['train loss', 'validation loss'], loc='upper right')
```

Expected output:

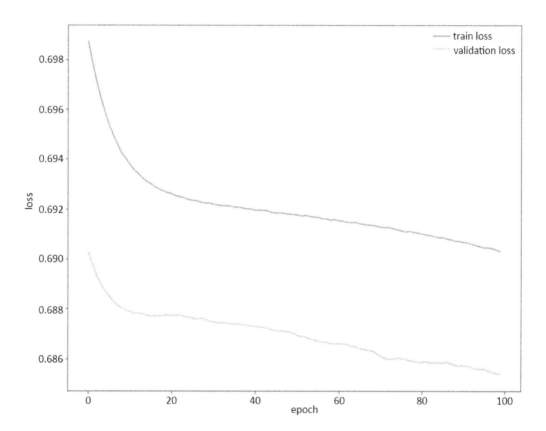

Figure 3.31: A plot of the training error rate and test error rate while training the model

7. Print the values of the best accuracy that was reached on the training set and on the test set, as well as the **loss** and **accuracy** that was evaluated on the **test** dataset.

```
print(f"Best Accuracy on training set = \
{max(history.history['accuracy'])*100:.3f}%")
print(f"Best Accuracy on validation set = \
{max(history.history['val_accuracy'])*100:.3f}%")

test_loss, test_acc = \
classifier.evaluate(X_test, y_test['AdvancedFibrosis'])

print(f'The loss on the test set is {test_loss:.4f} and \
the accuracy is {test_acc*100:.3f}%')
```

The following image shows the output of the preceding code:

```
Best Accuracy on training set = 52.959%
Best Accuracy on validation set = 58.559%
277/277 [==============================] - 0s 25us/step
The loss on the test set is 0.6885 and the accuracy is 55.235%
```

8. Implement a deep neural network with two hidden layers of sizes **4** and **2** with a **tanh activation function**, an output layer with one node, and a **sigmoid activation function**. Finally, compile the model and print out a summary of the model:

```
np.random.seed(seed)
random.set_seed(seed)

# define the keras model
classifier = Sequential()
classifier.add(Dense(units = 4, activation = 'tanh', \
                input_dim = X_train.shape[1]))
classifier.add(Dense(units = 2, activation = 'tanh'))
classifier.add(Dense(units = 1, activation = 'sigmoid'))
classifier.compile(optimizer = 'sgd', loss = 'binary_crossentropy', \
                metrics = ['accuracy'])

classifier.summary()
```

```
Model: "sequential_2"
```

Layer (type)	Output Shape	Param #
dense_3 (Dense)	(None, 4)	116
dense_4 (Dense)	(None, 2)	10
dense_5 (Dense)	(None, 1)	3

```
Total params: 129
Trainable params: 129
Non-trainable params: 0
```

Figure 3.32: The architecture of the neural network

9. Fit the model to the training data:

```
history=classifier.fit(X_train, y_train, batch_size = 20, \
                       epochs = 100, validation_split=0.1, \
                       shuffle=False)
```

10. Plot training and test error plots with two hidden layers of size 4 and 2. Print the best accuracy that was reached on the training and test sets:

```
# plot training error and test error plots
plt.plot(history.history['loss'])
plt.plot(history.history['val_loss'])
plt.ylabel('loss')
plt.xlabel('epoch')
plt.legend(['train loss', 'validation loss'], loc='upper right')
```

Expected output:

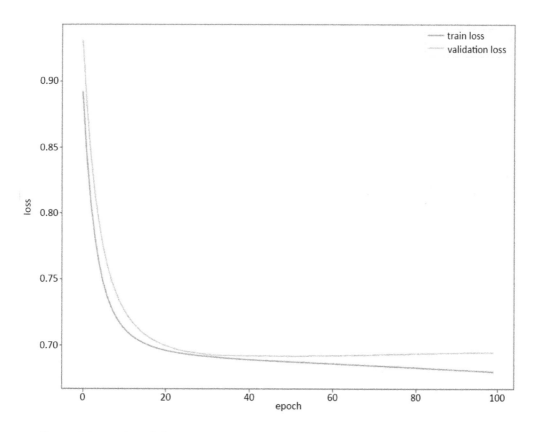

Figure 3.33: A plot of the training error and test error rates while training the model

11. Print the values of the best accuracy that was achieved on the **training** set and on the **test** set, as well as the **loss** and **accuracy** that was evaluated on the test dataset.

```
print(f"Best Accuracy on training set = \
{max(history.history['accuracy'])*100:.3f}%")
print(f"Best Accuracy on validation set = \
{max(history.history['val_accuracy'])*100:.3f}%")

test_loss, test_acc = \
classifier.evaluate(X_test, y_test['AdvancedFibrosis'])

print(f'The loss on the test set is {test_loss:.4f} and \
the accuracy is {test_acc*100:.3f}%')
```

The following shows the output of the preceding code:

```
Best Accuracy on training set = 57.272%
Best Accuracy on test set = 54.054%
277/277 [==============================] - 0s 41us/step
The loss on the test set is 0.7016 and the accuracy is 49.819%
```

> ## NOTE
>
> To access the source code for this specific section, please refer to https://packt.live/2BrIRMF.
>
> You can also run this example online at https://packt.live/2NUI22A.

CHAPTER 4: EVALUATING YOUR MODEL WITH CROSS-VALIDATION USING KERAS WRAPPERS

ACTIVITY 4.01: MODEL EVALUATION USING CROSS-VALIDATION FOR AN ADVANCED FIBROSIS DIAGNOSIS CLASSIFIER

In this activity, we are going to use what we learned in this topic to train and evaluate a deep learning model using **k-fold cross-validation**. We will use the model that resulted in the best test error rate from the previous activity and the goal will be to compare the cross-validation error rate with the training set/test set approach error rate. The dataset we will use is the hepatitis C dataset, in which we will build a classification model to predict which patients get advanced fibrosis. Follow these steps to complete this activity:

1. Load the dataset and print the number of records and features in the dataset, as well as the number of possible classes in the target dataset:

```
# Load the dataset
import pandas as pd
X = pd.read_csv('../data/HCV_feats.csv')
y = pd.read_csv('../data/HCV_target.csv')

# Print the sizes of the dataset
print("Number of Examples in the Dataset = ", X.shape[0])
print("Number of Features for each example = ", X.shape[1])
print("Possible Output Classes = ", \
        y['AdvancedFibrosis'].unique())
```

Here's the expected output:

```
Number of Examples in the Dataset = 1385
Number of Features for each example = 28
Possible Output Classes = [0 1]
```

2. Define the function that returns the Keras model. First, import the necessary libraries for Keras. Inside the function, instantiate the sequential model and add two dense layers, with the first of **size 4** and the second of **size 2**, both with **tanh activation** functions. Add the output layer with a **sigmoid activation** function. Compile the model and return the model from the function:

```
from keras.models import Sequential
from keras.layers import Dense
# Create the function that returns the keras model
def build_model():
    model = Sequential()
    model.add(Dense(4, input_dim=X.shape[1], activation='tanh'))
    model.add(Dense(2, activation='tanh'))
    model.add(Dense(1, activation='sigmoid'))
    model.compile(loss='binary_crossentropy', optimizer='adam', \
                  metrics=['accuracy'])
    return model
```

3. Scale the training data using the **StandardScaler** function. Set the seed so that the model is reproducible. Define the **n_folds**, **epochs**, and **batch_size** hyperparameters. Then, build the Keras wrapper with scikit-learn, define the **cross-validation** iterator, perform **k-fold cross-validation**, and store the scores:

```
# import required packages
import numpy as np
from tensorflow import random
from keras.wrappers.scikit_learn import KerasClassifier
from sklearn.model_selection import StratifiedKFold
from sklearn.model_selection import cross_val_score
from sklearn.preprocessing import StandardScaler

sc = StandardScaler()
X = pd.DataFrame(sc.fit_transform(X), columns=X.columns)
"""
define a seed for random number generator so the result will be
reproducible
"""
seed = 1
np.random.seed(seed)
random.set_seed(seed)
```

```
"""
determine the number of folds for k-fold cross-validation, number of
epochs and batch size
"""
n_folds = 5
epochs = 100
batch_size = 20


# build the scikit-learn interface for the keras model
classifier = KerasClassifier(build_fn=build_model, \
                             epochs=epochs, \
                             batch_size=batch_size, \
                             verbose=1, shuffle=False)
# define the cross-validation iterator
kfold = StratifiedKFold(n_splits=n_folds, shuffle=True, \
                 random_state=seed)
"""
perform the k-fold cross-validation and store the scores in results
"""
results = cross_val_score(classifier, X, y, cv=kfold)
```

4. For each of the folds, print the accuracy stored in the **results** parameter:

```
# print accuracy for each fold
for f in range(n_folds):
    print("Test accuracy at fold ", f+1, " = ", results[f])
print("\n")


"""
print overall cross-validation accuracy plus the standard deviation
of the accuracies
"""
print("Final Cross-validation Test Accuracy:", results.mean())
print("Standard Deviation of Final Test Accuracy:", results.std())
```

Here's the expected output:

```
Test accuracy at fold 1 = 0.5198556184768677
Test accuracy at fold 2 = 0.4693140685558319
Test accuracy at fold 3 = 0.512635350227356
Test accuracy at fold 4 = 0.5740072131156921
Test accuracy at fold 5 = 0.5523465871810913


Final Cross-Validation Test Accuracy: 0.5256317675113678
Standard Deviation of Final Test Accuracy: 0.03584760640500936
```

> **NOTE**
>
> To access the source code for this specific section, please refer to
> https://packt.live/3eWgR2b.
>
> You can also run this example online at https://packt.live/3iBYtOi.

ACTIVITY 4.02: MODEL SELECTION USING CROSS-VALIDATION FOR THE ADVANCED FIBROSIS DIAGNOSIS CLASSIFIER

In this activity, we are going to improve our classifier for the hepatitis C dataset by using cross-validation for model selection and hyperparameter selection. Follow these steps to complete this activity:

1. Import all the required packages and load the dataset. Scale the dataset using the **StandardScaler** function:

```
# import the required packages
from keras.models import Sequential
from keras.layers import Dense
from keras.wrappers.scikit_learn import KerasClassifier
from sklearn.model_selection import StratifiedKFold
from sklearn.model_selection import cross_val_score
import numpy as np
import pandas as pd
from sklearn.preprocessing import StandardScaler
from tensorflow import random

# Load the dataset
```

```
X = pd.read_csv('../data/HCV_feats.csv')
y = pd.read_csv('../data/HCV_target.csv')

sc = StandardScaler()
X = pd.DataFrame(sc.fit_transform(X), columns=X.columns)
```

2. Define three functions, each returning a different Keras model. The first model should have three hidden layers of **size** 4, the second model should have two hidden layers, the first of **size** 4 and the second of **size** 2, and the third model should have two hidden layers of **size** 8. Use function parameters for the activation functions and optimizers so that they can be passed through to the model. The goal is to find out which of these three models leads to the lowest cross-validation error rate:

```
# Create the function that returns the keras model 1
def build_model_1(activation='relu', optimizer='adam'):
    # create model 1
    model = Sequential()
    model.add(Dense(4, input_dim=X.shape[1], \
                    activation=activation))
    model.add(Dense(4, activation=activation))
    model.add(Dense(4, activation=activation))
    model.add(Dense(1, activation='sigmoid'))
    # Compile model
    model.compile(loss='binary_crossentropy', \
                  optimizer=optimizer, metrics=['accuracy'])
    return model


# Create the function that returns the keras model 2
def build_model_2(activation='relu', optimizer='adam'):
    # create model 2
    model = Sequential()
    model.add(Dense(4, input_dim=X.shape[1], \
                    activation=activation))
    model.add(Dense(2, activation=activation))
    model.add(Dense(1, activation='sigmoid'))
    # Compile model
    model.compile(loss='binary_crossentropy', \
                  optimizer=optimizer, metrics=['accuracy'])
    return model
```

```
# Create the function that returns the keras model 3
def build_model_3(activation='relu', optimizer='adam'):
    # create model 3
    model = Sequential()
    model.add(Dense(8, input_dim=X.shape[1], \
                    activation=activation))
    model.add(Dense(8, activation=activation))
    model.add(Dense(1, activation='sigmoid'))
    # Compile model
    model.compile(loss='binary_crossentropy', \
                  optimizer=optimizer, metrics=['accuracy'])
    return model
```

Write the code that will loop over the three models and perform **5-fold cross-validation**. Set the seed so that the models are reproducible and define the **n_folds**, **batch_size**, and **epochs** hyperparameters. Store the results from applying the **cross_val_score** function when training the models:

```
"""
define a seed for random number generator so the result will be
reproducible
"""
seed = 2
np.random.seed(seed)
random.set_seed(seed)
"""
determine the number of folds for k-fold cross-validation, number of
epochs and batch size
"""
n_folds = 5
batch_size=20
epochs=100

# define the list to store cross-validation scores
results_1 = []

# define the possible options for the model
models = [build_model_1, build_model_2, build_model_3]

# loop over models
for m in range(len(models)):
```

```
        # build the scikit-learn interface for the keras model
        classifier = KerasClassifier(build_fn=models[m], \
                                     epochs=epochs, \
                                     batch_size=batch_size, \
                                     verbose=0, shuffle=False)
        # define the cross-validation iterator
        kfold = StratifiedKFold(n_splits=n_folds, shuffle=True, \
                                random_state=seed)
        """
        perform the k-fold cross-validation and store the scores
        in result
        """

        result = cross_val_score(classifier, X, y, cv=kfold)
        # add the scores to the results list
        results_1.append(result)

# Print cross-validation score for each model
for m in range(len(models)):
    print("Model", m+1,"Test Accuracy =", results_1[m].mean())
```

Here's an example output. In this instance, **Model 2** has the best cross-validation test accuracy, as you can see below:

```
Model 1 Test Accuracy = 0.4996389865875244
Model 2 Test Accuracy = 0.5148014307022095
Model 3 Test Accuracy = 0.5097472846508027
```

3. Choose the model with the highest accuracy score and repeat *step 2* by iterating over the **epochs = [100, 200]** and **batches = [10, 20]** values and performing **5-fold cross-validation**:

```
"""
define a seed for random number generator so the result will be
reproducible
"""
np.random.seed(seed)
random.set_seed(seed)
# determine the number of folds for k-fold cross-validation
n_folds = 5
# define possible options for epochs and batch_size
epochs = [100, 200]
batches = [10, 20]
```

```
# define the list to store cross-validation scores
results_2 = []
# loop over all possible pairs of epochs, batch_size
for e in range(len(epochs)):
    for b in range(len(batches)):
        # build the scikit-learn interface for the keras model
        classifier = KerasClassifier(build_fn=build_model_2, \
                                     epochs=epochs[e], \
                                     batch_size=batches[b], \
                                     verbose=0)
        # define the cross-validation iterator
        kfold = StratifiedKFold(n_splits=n_folds, shuffle=True, \
                        random_state=seed)
        # perform the k-fold cross-validation.
        # store the scores in result
        result = cross_val_score(classifier, X, y, cv=kfold)
        # add the scores to the results list
        results_2.append(result)


"""
Print cross-validation score for each possible pair of epochs, batch_
size
"""
c = 0
for e in range(len(epochs)):
    for b in range(len(batches)):
        print("batch_size =", batches[b],", epochs =", epochs[e], \
            ", Test Accuracy =", results_2[c].mean())
        c += 1
```

Here's an example output:

```
batch_size = 10 , epochs = 100 , Test Accuracy = 0.5010830342769623
batch_size = 20 , epochs = 100 , Test Accuracy = 0.5126353740692139
batch_size = 10 , epochs = 200 , Test Accuracy = 0.5176895320416497
batch_size = 20 , epochs = 200 , Test Accuracy = 0.5075812220573426
```

In this case, the **batch_size= 10**, **epochs=200** pair has the best cross-validation test accuracy.

4. Choose the batch size and epochs with the highest accuracy score and repeat *step 3* by iterating over the **optimizers = ['rmsprop', 'adam','sgd']** and **activations = ['relu', 'tanh']** values and performing **5-fold cross-validation**:

```
"""
define a seed for random number generator so the result will be
reproducible
"""
np.random.seed(seed)
random.set_seed(seed)
"""
determine the number of folds for k-fold cross-validation, number of
epochs and batch size
"""
n_folds = 5
batch_size = 10
epochs = 200
# define the list to store cross-validation scores
results_3 = []
# define possible options for optimizer and activation
optimizers = ['rmsprop', 'adam','sgd']
activations = ['relu', 'tanh']
# loop over all possible pairs of optimizer, activation
for o in range(len(optimizers)):
    for a in range(len(activations)):
        optimizer = optimizers[o]
        activation = activations[a]
        # build the scikit-learn interface for the keras model
        classifier = KerasClassifier(build_fn=build_model_2, \
                                     epochs=epochs, \
                                     batch_size=batch_size, \
                                     verbose=0, shuffle=False)
        # define the cross-validation iterator
        kfold = StratifiedKFold(n_splits=n_folds, shuffle=True, \
                                random_state=seed)
        # perform the k-fold cross-validation.
        # store the scores in result
        result = cross_val_score(classifier, X, y, cv=kfold)
        # add the scores to the results list
        results_3.append(result)
```

```
"""
Print cross-validation score for each possible pair of optimizer,
activation
"""
c = 0
for o in range(len(optimizers)):
    for a in range(len(activations)):
        print("activation = ", activations[a],", optimizer = ", \
            optimizers[o], ", Test accuracy = ", \
            results_3[c].mean())
        c += 1
```

Here's the expected output:

```
activation =  relu , optimizer =  rmsprop ,
Test accuracy =  0.5234657049179077
activation =  tanh , optimizer =  rmsprop ,
Test accuracy =  0.49602887630462644
activation =  relu , optimizer =  adam ,
Test accuracy =  0.5039711117744445
activation =  tanh , optimizer =  adam ,
Test accuracy =  0.4989169597625732
activation =  relu , optimizer =  sgd ,
Test accuracy =  0.48953068256378174
activation =  tanh , optimizer =  sgd ,
Test accuracy =  0.5191335678100586
```

Here, the **activation='relu'** and **optimizer='rmsprop'** pair has the best cross-validation test accuracy. Also, the **activation='tanh'** and **optimizer='sgd'** pair results in the second-best performance.

> ### NOTE
>
> To access the source code for this specific section, please refer to https://packt.live/2D3AlhD.
>
> You can also run this example online at https://packt.live/2NUpiiC.

ACTIVITY 4.03: MODEL SELECTION USING CROSS-VALIDATION ON A TRAFFIC VOLUME DATASET

In this activity, you are going to practice model selection using cross-validation one more time. Here, we are going to use a simulated dataset that represents a target variable representing the volume of traffic in cars/hour across a city bridge and various normalized features related to traffic data such as time of day and the traffic volume on the previous day. Our goal is to build a model that predicts the traffic volume across the city bridge given the various features. Follow these steps to complete this activity:

1. Import all the required packages and load the dataset:

```
# import the required packages
from keras.models import Sequential
from keras.layers import Dense
from keras.wrappers.scikit_learn import KerasRegressor
from sklearn.model_selection import KFold
from sklearn.model_selection import cross_val_score
from sklearn.preprocessing import StandardScaler
from sklearn.pipeline import make_pipeline
import numpy as np
import pandas as pd
from tensorflow import random
```

2. Load the dataset, print the input and output size for the feature dataset, and print the possible classes in the target dataset. Also, print the range of the output:

```
# Load the dataset
# Load the dataset
X = pd.read_csv('../data/traffic_volume_feats.csv')
y = pd.read_csv('../data/traffic_volume_target.csv')
# Print the sizes of input data and output data
print("Input data size = ", X.shape)
print("Output size = ", y.shape)
# Print the range for output
print(f"Output Range = ({y['Volume'].min()}, \
{ y['Volume'].max()})")
```

Here's the expected output:

```
Input data size =  (10000, 10)
Output size =  (10000, 1)
Output Range = (0.000000, 584.000000)
```

3. Define three functions, each returning a different Keras model. The first model should have one hidden layer of **size 10**, the second model should have two hidden layers of **size 10**, and the third model should have three hidden layers of **size 10**. Use function parameters for the optimizers so that they can be passed through to the model. The goal is to find out which of these three models leads to the lowest cross-validation error rate:

```
# Create the function that returns the keras model 1
def build_model_1(optimizer='adam'):
    # create model 1
    model = Sequential()
    model.add(Dense(10, input_dim=X.shape[1], activation='relu'))
    model.add(Dense(1))
    # Compile model
    model.compile(loss='mean_squared_error', optimizer=optimizer)
    return model


# Create the function that returns the keras model 2
def build_model_2(optimizer='adam'):
    # create model 2
    model = Sequential()
    model.add(Dense(10, input_dim=X.shape[1], activation='relu'))
    model.add(Dense(10, activation='relu'))
    model.add(Dense(1))
    # Compile model
    model.compile(loss='mean_squared_error', optimizer=optimizer)
    return model


# Create the function that returns the keras model 3
def build_model_3(optimizer='adam'):
    # create model 3
    model = Sequential()
    model.add(Dense(10, input_dim=X.shape[1], activation='relu'))
    model.add(Dense(10, activation='relu'))
```

```
    model.add(Dense(10, activation='relu'))
    model.add(Dense(1))
    # Compile model
    model.compile(loss='mean_squared_error', optimizer=optimizer)
    return model
```

4. Write the code that will loop over the three models and perform **5-fold cross-validation**. Set the seed so that the models are reproducible and define the **n_folds** hyperparameters. Store the results from applying the **cross_val_score** function when training the models:

```
"""
define a seed for random number generator so the result will be
reproducible
"""
seed = 1
np.random.seed(seed)
random.set_seed(seed)
# determine the number of folds for k-fold cross-validation
n_folds = 5
# define the list to store cross-validation scores
results_1 = []
# define the possible options for the model
models = [build_model_1, build_model_2, build_model_3]
# loop over models
for i in range(len(models)):
    # build the scikit-learn interface for the keras model
    regressor = KerasRegressor(build_fn=models[i], epochs=100, \
                               batch_size=50, verbose=0, \
                               shuffle=False)
    """
    build the pipeline of transformations so for each fold training
    set will be scaled and test set will be scaled accordingly.
    """
    model = make_pipeline(StandardScaler(), regressor)
    # define the cross-validation iterator
    kfold = KFold(n_splits=n_folds, shuffle=True, \
                  random_state=seed)
    # perform the k-fold cross-validation.
```

```
        # store the scores in result
        result = cross_val_score(model, X, y, cv=kfold)
        # add the scores to the results list
        results_1.append(result)

# Print cross-validation score for each model
for i in range(len(models)):
    print("Model ", i+1," test error rate = ", \
            abs(results_1[i].mean()))
```

The following is the expected output:

```
Model  1  test error rate =  25.48777518749237
Model  2  test error rate =  25.30460816860199
Model  3  test error rate =  25.390239462852474
```

Model 2 (a two-layer neural network) has the lowest test error rate.

5. Choose the model with the lowest test error rate and repeat *step 4* while iterating over **epochs = [80, 100]** and **batches = [50, 25]** and performing **5-fold cross-validation**:

```
"""
define a seed for random number generator so the result will be
reproducible
"""
np.random.seed(seed)
random.set_seed(seed)
# determine the number of folds for k-fold cross-validation
n_folds = 5
# define the list to store cross-validation scores
results_2 = []
# define possible options for epochs and batch_size
epochs = [80, 100]
batches = [50, 25]
# loop over all possible pairs of epochs, batch_size
for i in range(len(epochs)):
    for j in range(len(batches)):
        # build the scikit-learn interface for the keras model
        regressor = KerasRegressor(build_fn=build_model_2, \
                                   epochs=epochs[i], \
                                   batch_size=batches[j], \
                                   verbose=0, shuffle=False)
```

```
        """
        build the pipeline of transformations so for each fold
        training set will be scaled and test set will be scaled
        accordingly.
        """
        model = make_pipeline(StandardScaler(), regressor)
        # define the cross-validation iterator
        kfold = KFold(n_splits=n_folds, shuffle=True, \
                      random_state=seed)
        # perform the k-fold cross-validation.
        # store the scores in result
        result = cross_val_score(model, X, y, cv=kfold)
        # add the scores to the results list
        results_2.append(result)


"""
Print cross-validation score for each possible pair of epochs, batch_
size
"""
c = 0
for i in range(len(epochs)):
    for j in range(len(batches)):
        print("batch_size = ", batches[j],\
              ", epochs = ", epochs[i], \
              ", Test error rate = ", abs(results_2[c].mean()))
        c += 1
```

Here's the expected output:

```
batch_size = 50 , epochs = 80 , Test error rate = 25.270704221725463
batch_size = 25 , epochs = 80 , Test error rate = 25.309741401672362
batch_size = 50 , epochs = 100 , Test error rate = 25.095393986701964
batch_size = 25 , epochs = 100 , Test error rate = 25.24592453837395
```

The **batch_size=5** and **epochs=100** pair has the lowest test error rate.

6. Choose the model with the highest accuracy score and repeat *step 2* by iterating over **optimizers = ['rmsprop', 'sgd', 'adam']** and performing **5-fold cross-validation**:

```
"""
define a seed for random number generator so the result will be
reproducible
"""
np.random.seed(seed)
random.set_seed(seed)
# determine the number of folds for k-fold cross-validation
n_folds = 5
# define the list to store cross-validation scores
results_3 = []
# define the possible options for the optimizer
optimizers = ['adam', 'sgd', 'rmsprop']
# loop over optimizers
for i in range(len(optimizers)):
    optimizer=optimizers[i]
    # build the scikit-learn interface for the keras model
    regressor = KerasRegressor(build_fn=build_model_2, \
                               epochs=100, batch_size=50, \
                               verbose=0, shuffle=False)
    """
    build the pipeline of transformations so for each fold training
    set will be scaled and test set will be scaled accordingly.
    """
    model = make_pipeline(StandardScaler(), regressor)
    # define the cross-validation iterator
    kfold = KFold(n_splits=n_folds, shuffle=True, \
                  random_state=seed)
    # perform the k-fold cross-validation.
    # store the scores in result
    result = cross_val_score(model, X, y, cv=kfold)
    # add the scores to the results list
    results_3.append(result)
# Print cross-validation score for each optimizer
```

```
for i in range(len(optimizers)):
    print("optimizer=", optimizers[i]," test error rate = ", \
        abs(results_3[i].mean()))
```

Here's the expected output:

```
optimizer= adam   test error rate =   25.391812739372256
optimizer= sgd    test error rate =   25.140230269432067
optimizer= rmsprop  test error rate =   25.217947859764102
```

optimizer='sgd' has the lowest test error rate, so we should proceed with this particular model.

> **NOTE**
>
> To access the source code for this specific section, please refer to https://packt.live/31TcYaD.
>
> You can also run this example online at https://packt.live/3iq6iqb.

CHAPTER 5: IMPROVING MODEL ACCURACY

ACTIVITY 5.01: WEIGHT REGULARIZATION ON AN AVILA PATTERN CLASSIFIER

In this activity, you will build a Keras model to perform classification on the Avila pattern dataset according to given network architecture and hyperparameter values. The goal is to apply different types of weight regularization on the model, that is, **L1** and **L2**, and observe how each type changes the result. Follow these steps to complete this activity:

1. Load the dataset and split the dataset into a **training set** and a **test set**:

```
# Load the dataset
import pandas as pd
X = pd.read_csv('../data/avila-tr_feats.csv')
y = pd.read_csv('../data/avila-tr_target.csv')


"""
Split the dataset into training set and test set with a 0.8-0.2 ratio
"""
from sklearn.model_selection import train_test_split
seed = 1
X_train, X_test, y_train, y_test = \
train_test_split(X, y, test_size=0.2, random_state=seed)
```

2. Define a Keras sequential model with three hidden layers, the first of **size 10**, the second of **size 6**, and the third of **size 4**. Finally, compile the model:

```
"""
define a seed for random number generator so the result will be
reproducible
"""
import numpy as np
from tensorflow import random
np.random.seed(seed)
random.set_seed(seed)

# define the keras model
from keras.models import Sequential
```

```
from keras.layers import Dense
model_1 = Sequential()
model_1.add(Dense(10, input_dim=X_train.shape[1], \
                  activation='relu'))
model_1.add(Dense(6, activation='relu'))
model_1.add(Dense(4, activation='relu'))
model_1.add(Dense(1, activation='sigmoid'))
model_1.compile(loss='binary_crossentropy', optimizer='sgd', \
                metrics=['accuracy'])
```

3. Fit the model to the training data to perform the classification, saving the results of the training process:

```
history=model_1.fit(X_train, y_train, batch_size = 20, epochs = 100, \
                    validation_data=(X_test, y_test), \
                    verbose=0, shuffle=False)
```

4. Plot the trends in training error and test error by importing the necessary libraries for plotting the loss and validation loss and saving them in the variable that was created when the model was fit to the training process. Print out the maximum validation accuracy:

```
import matplotlib.pyplot as plt
import matplotlib
%matplotlib inline

# plot training error and test error
matplotlib.rcParams['figure.figsize'] = (10.0, 8.0)
plt.plot(history.history['loss'])
plt.plot(history.history['val_loss'])
plt.ylim(0,1)
plt.ylabel('loss')
plt.xlabel('epoch')
plt.legend(['train loss', 'validation loss'], loc='upper right')

# print the best accuracy reached on the test set
print("Best Accuracy on Validation Set =", \
      max(history.history['val_accuracy']))
```

The following is the expected output:

```
Best Accuracy on Validation Set = 0.8024927973747253
```

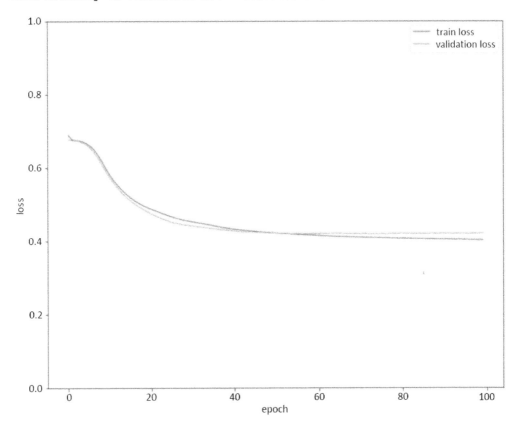

Figure 5.13: A plot of the training error and validation error during training for the model without regularization

The validation loss keeps decreasing along with the training loss. Despite having no regularization, this is a fairly good example of the training process since the bias and variance are fairly low.

5. Redefine the model, adding **L2 regularizers** with **lambda=0.01** to each hidden layer of the model. Repeat *steps 3* and *4* to train the model and plot the **training error** and **validation error**:

```
"""
set up a seed for random number generator so the result will be
reproducible
"""
```

```
np.random.seed(seed)
random.set_seed(seed)
# define the keras model with l2 regularization with lambda = 0.01
from keras.regularizers import l2
l2_param = 0.01
model_2 = Sequential()
model_2.add(Dense(10, input_dim=X_train.shape[1], \
                activation='relu', \
                kernel_regularizer=l2(l2_param)))
model_2.add(Dense(6, activation='relu', \
                kernel_regularizer=l2(l2_param)))
model_2.add(Dense(4, activation='relu', \
                kernel_regularizer=l2(l2_param)))
model_2.add(Dense(1, activation='sigmoid'))
model_2.compile(loss='binary_crossentropy', optimizer='sgd', \
              metrics=['accuracy'])

# train the model using training set while evaluating on test set
history=model_2.fit(X_train, y_train, batch_size = 20, epochs = 100, \
                  validation_data=(X_test, y_test), \
                  verbose=0, shuffle=False)

plt.plot(history.history['loss'])
plt.plot(history.history['val_loss'])
plt.ylim(0,1)
plt.ylabel('loss')
plt.xlabel('epoch')
plt.legend(['train loss', 'validation loss'], loc='upper right')
# print the best accuracy reached on the test set
print("Best Accuracy on Validation Set =", \
    max(history.history['val_accuracy']))
```

The following is the expected output:

```
Best Accuracy on Validation Set = 0.797698974609375
```

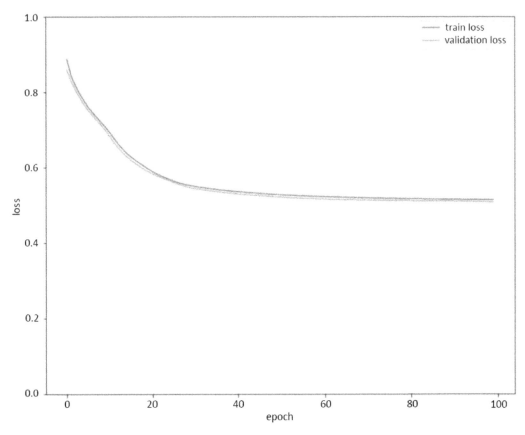

**Figure 5.14: A plot of the training error and validation error during training
for the model with L2 weight regularization (lambda=0.01)**

As shown from the preceding plots, the test error almost plateaus after being decreased to a certain amount. The gap between the training error and the validation error at the end of the training process (the bias) is slightly smaller, which is indicative of reduced overfitting of the model for the training examples.

6. Repeat the previous step with **lambda=0.1** for the **L2 parameter**—redefine the model with the new lambda parameter, fit the model to the training data, and repeat *step 4* to plot the training error and validation error:

```
"""
set up a seed for random number generator so the result will be
reproducible
"""
```

```python
np.random.seed(seed)
random.set_seed(seed)
from keras.regularizers import l2
l2_param = 0.1
model_3 = Sequential()
model_3.add(Dense(10, input_dim=X_train.shape[1], \
                  activation='relu', \
                  kernel_regularizer=l2(l2_param)))

model_3.add(Dense(6, activation='relu', \
                  kernel_regularizer=l2(l2_param)))

model_3.add(Dense(4, activation='relu', \
                  kernel_regularizer=l2(l2_param)))
model_3.add(Dense(1, activation='sigmoid'))
model_3.compile(loss='binary_crossentropy', optimizer='sgd', \
                metrics=['accuracy'])

# train the model using training set while evaluating on test set
history=model_3.fit(X_train, y_train, batch_size = 20, \
                    epochs = 100, validation_data=(X_test, y_test), \
                    verbose=0, shuffle=False)

# plot training error and test error
matplotlib.rcParams['figure.figsize'] = (10.0, 8.0)
plt.plot(history.history['loss'])
plt.plot(history.history['val_loss'])
plt.ylim(0,1)
plt.ylabel('loss')
plt.xlabel('epoch')
plt.legend(['train loss', 'validation loss'], loc='upper right')
# print the best accuracy reached on the test set
print("Best Accuracy on Validation Set =", \
      max(history.history['val_accuracy']))
```

The following is the expected output:

```
Best Accuracy on Validation Set = 0.5910834074020386
```

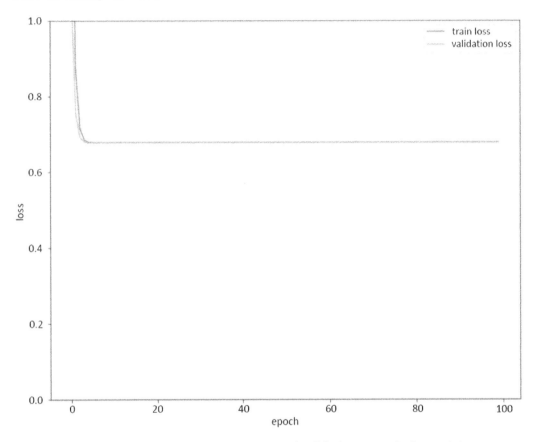

Figure 5.15: A plot of the training error and validation error during training
for the model with L2 weight regularization (lambda=0.1)

The training and validation error quickly plateau and are much higher than they were for the models we created with a lower **L2 parameter**, indicating that we have penalized the model so much that it has not had the flexibility to learn the underlying function of the training data. Following this, we will reduce the value of the regularization parameter to prevent it from penalizing the model as much.

7. Repeat the previous step, this time with **lambda=0.005**. Repeat *step 4* to plot the training error and validation error:

```
"""
set up a seed for random number generator so the result will be
reproducible
"""
```

```python
np.random.seed(seed)
random.set_seed(seed)


# define the keras model with l2 regularization with lambda = 0.05
from keras.regularizers import l2
l2_param = 0.005
model_4 = Sequential()
model_4.add(Dense(10, input_dim=X_train.shape[1], \
                  activation='relu', \
                  kernel_regularizer=l2(l2_param)))
model_4.add(Dense(6, activation='relu', \
                  kernel_regularizer=l2(l2_param)))
model_4.add(Dense(4, activation='relu', \
                  kernel_regularizer=l2(l2_param)))
model_4.add(Dense(1, activation='sigmoid'))
model_4.compile(loss='binary_crossentropy', optimizer='sgd', \
                metrics=['accuracy'])


# train the model using training set while evaluating on test set
history=model_4.fit(X_train, y_train, batch_size = 20, \
                    epochs = 100, validation_data=(X_test, y_test), \
                    verbose=0, shuffle=False)


# plot training error and test error
matplotlib.rcParams['figure.figsize'] = (10.0, 8.0)
plt.plot(history.history['loss'])
plt.plot(history.history['val_loss'])
plt.ylim(0,1)
plt.ylabel('loss')
plt.xlabel('epoch')
plt.legend(['train loss', 'validation loss'], loc='upper right')
# print the best accuracy reached on the test set
print("Best Accuracy on Validation Set =", \
      max(history.history['val_accuracy']))
```

The following is the expected output:

```
Best Accuracy on Validation Set = 0.8024927973747253
```

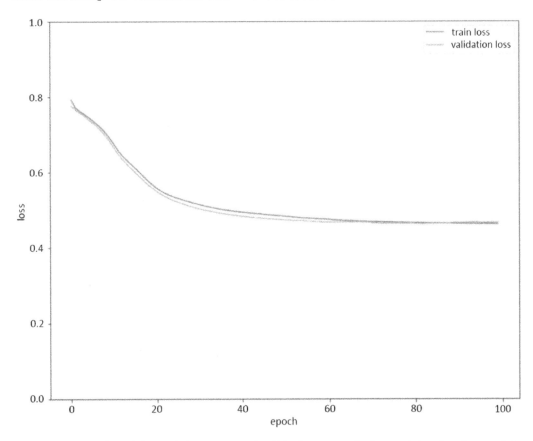

Figure 5.16: A plot of the training error and validation error during training for the model with L2 weight regularization (lambda=0.005)

The value for the **L2 weight** regularization achieves the highest accuracy that was evaluated on the validation data of all the models with **L2 regularization**, but it is slightly lower than without regularization. Again, the test error does not increase a significant amount after being decreased to a certain value, which is indicative of the model not overfitting the training examples. It seems that **L2 weight regularization** with **lambda=0.005** achieves the lowest validation error while preventing the model from overfitting.

8. Add **L1 regularizers** with **lambda=0.01** to the hidden layers of your model. Redefine the model with the new lambda parameter, fit the model to the training data, and repeat *step 4* to plot the training error and validation error:

```
"""
set up a seed for random number generator so the result will be
reproducible
"""
np.random.seed(seed)
random.set_seed(seed)

# define the keras model with l1 regularization with lambda = 0.01
from keras.regularizers import l1
l1_param = 0.01
model_5 = Sequential()
model_5.add(Dense(10, input_dim=X_train.shape[1], \
                  activation='relu', \
                  kernel_regularizer=l1(l1_param)))
model_5.add(Dense(6, activation='relu', \
                  kernel_regularizer=l1(l1_param)))
model_5.add(Dense(4, activation='relu', \
                  kernel_regularizer=l1(l1_param)))
model_5.add(Dense(1, activation='sigmoid'))
model_5.compile(loss='binary_crossentropy', optimizer='sgd', \
                metrics=['accuracy'])

# train the model using training set while evaluating on test set
history=model_5.fit(X_train, y_train, batch_size = 20, \
                    epochs = 100, validation_data=(X_test, y_test), \
                    verbose=0, shuffle=True)

# plot training error and test error
matplotlib.rcParams['figure.figsize'] = (10.0, 8.0)
plt.plot(history.history['loss'])
plt.plot(history.history['val_loss'])
plt.ylim(0,1)
plt.ylabel('loss')
```

```
plt.xlabel('epoch')
plt.legend(['train loss', 'validation loss'], loc='upper right')
# print the best accuracy reached on the test set
print("Best Accuracy on Validation Set =", \
    max(history.history['val_accuracy']))
```

The following is the expected output:

```
Best Accuracy on Validation Set = 0.5910834074020386
```

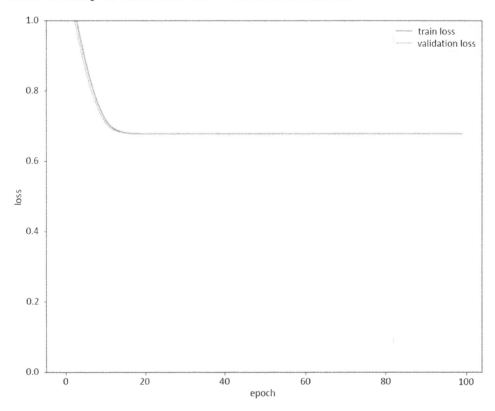

Figure 5.17: A plot of the training error and validation error during training
for the model with L1 weight regularization (lambda=0.01)

9. Repeat the previous step with **lambda=0.005** for the **L1 parameter**—
   redefine the model with the new lambda parameter, fit the model to
   the training data, and repeat *step 4* to plot the **training error** and
   validation error:

```
"""
set up a seed for random number generator so the result will be
reproducible
"""
```

```
np.random.seed(seed)
random.set_seed(seed)

# define the keras model with l1 regularization with lambda = 0.1
from keras.regularizers import l1
l1_param = 0.005
model_6 = Sequential()
model_6.add(Dense(10, input_dim=X_train.shape[1], \
                  activation='relu', \
                  kernel_regularizer=l1(l1_param)))
model_6.add(Dense(6, activation='relu', \
                  kernel_regularizer=l1(l1_param)))
model_6.add(Dense(4, activation='relu', \
                  kernel_regularizer=l1(l1_param)))
model_6.add(Dense(1, activation='sigmoid'))
model_6.compile(loss='binary_crossentropy', optimizer='sgd', \
                metrics=['accuracy'])

# train the model using training set while evaluating on test set
history=model_6.fit(X_train, y_train, batch_size = 20, \
                    epochs = 100, validation_data=(X_test, y_test), \
                    verbose=0, shuffle=False)

# plot training error and test error
matplotlib.rcParams['figure.figsize'] = (10.0, 8.0)
plt.plot(history.history['loss'])
plt.plot(history.history['val_loss'])
plt.ylim(0,1)
plt.ylabel('loss')
plt.xlabel('epoch')
plt.legend(['train loss', 'validation loss'], loc='upper right')
# print the best accuracy reached on the test set
print("Best Accuracy on Validation Set =", \
      max(history.history['val_accuracy']))
```

The following is the expected output:

```
Best Accuracy on Validation Set = 0.7794822454452515
```

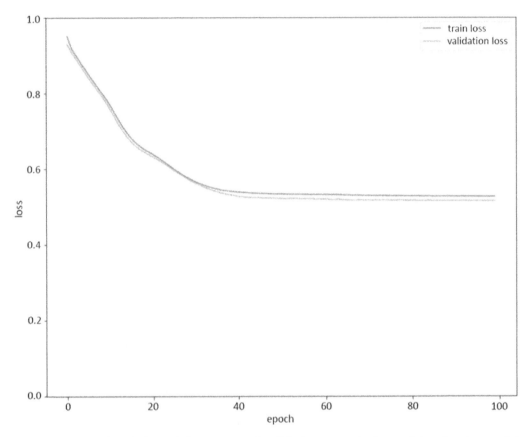

Figure 5.18: The plot of the training error and validation error during training for the model with L1 weight regularization (lambda=0.005)

It seems that **L1 weight regularization** with **lambda=0.005** achieves a better test error while preventing the model from overfitting since the value of **lambda=0.01** is too restrictive and prevents the model from learning the underlying function of the training data.

10. Add **L1** and **L2 regularizers** with an **L1** of **lambda=0.005** and an **L2** of **lambda = 0.005** to the hidden layers of your model. Then, repeat *step 4* to plot the training error and validation error:

```
"""
set up a seed for random number generator so the result will be
reproducible
"""
```

```python
np.random.seed(seed)
random.set_seed(seed)


"""
define the keras model with l1_l2 regularization with l1_lambda =
0.005 and l2_lambda = 0.005
"""

from keras.regularizers import l1_l2
l1_param = 0.005
l2_param = 0.005
model_7 = Sequential()
model_7.add(Dense(10, input_dim=X_train.shape[1], \
            activation='relu', \
            kernel_regularizer=l1_l2(l1=l1_param, l2=l2_param)))
model_7.add(Dense(6, activation='relu', \
                  kernel_regularizer=l1_l2(l1=l1_param, \
                                           l2=l2_param)))
model_7.add(Dense(4, activation='relu', \
                  kernel_regularizer=l1_l2(l1=l1_param, \
                                           l2=l2_param)))
model_7.add(Dense(1, activation='sigmoid'))
model_7.compile(loss='binary_crossentropy', optimizer='sgd', \
                metrics=['accuracy'])

# train the model using training set while evaluating on test set
history=model_7.fit(X_train, y_train, batch_size = 20, \
                    epochs = 100, validation_data=(X_test, y_test), \
                    verbose=0, shuffle=True)

# plot training error and test error
matplotlib.rcParams['figure.figsize'] = (10.0, 8.0)
plt.plot(history.history['loss'])
plt.plot(history.history['val_loss'])
plt.ylim(0,1)
plt.ylabel('loss')
plt.xlabel('epoch')
plt.legend(['train loss', 'validation loss'], loc='upper right')
```

```
# print the best accuracy reached on the test set
print("Best Accuracy on Validation Set =", \
      max(history.history['val_accuracy']))
```

The following is the expected output:

```
Best Accuracy on Validation Set = 0.5925215482711792
```

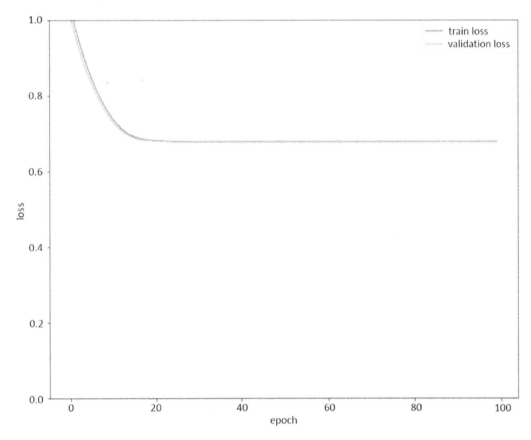

Figure 5.19: A plot of the training error and validation error during training for the model with L1 lambda equal to 0.005 and L2 lambda equal to 0.005

While **L1** and **L2 regularization** are successful in preventing the model from overfitting, the variance in the model is very low. However, the accuracy that's obtained on the validation data is not as high as the model that was trained with no regularization or the model that was trained with the **L2 regularization lambda=0.005** or **L1 regularization lambda=0.005** parameters individually.

> **NOTE**
>
> To access the source code for this specific section, please refer to
> https://packt.live/31BUf34.
>
> You can also run this example online at https://packt.live/38n291s.

ACTIVITY 5.02: DROPOUT REGULARIZATION ON THE TRAFFIC VOLUME DATASET

In this activity, you will start with the model from *Activity 4.03, Model Selection Using Cross-Validation on a Traffic Volume Dataset*, of *Chapter 4, Evaluating Your Model with Cross-Validation Using Keras Wrappers*. You will use the training set/test set approach to train and evaluate the model, plot the trends in training error and the generalization error, and observe the model overfitting the data examples. Then, you will attempt to improve model performance by addressing the overfitting issue through the use of dropout regularization. In particular, you will try to find out which layers you should add dropout regularization to and what **rate** value will improve this specific model the most. Follow these steps to complete this exercise:

1. Load the dataset using the pandas **read_csv** function, split the dataset into a training set and test set into an **80-20** ratio using **train_test_split**, and scale the input data using **StandardScaler**:

```
# Load the dataset
import pandas as pd
X = pd.read_csv('../data/traffic_volume_feats.csv')
y = pd.read_csv('../data/traffic_volume_target.csv')

"""
Split the dataset into training set and test set with an 80-20 ratio
"""
from sklearn.model_selection import train_test_split
seed=1
X_train, X_test, y_train, y_test = \
train_test_split(X, y, test_size=0.2, random_state=seed)
```

2. Set a seed so that the model can be reproduced. Next, define a Keras sequential model with two hidden layers of **size 10**, both with **ReLU activation** functions. Add an output layer with no activation function and compile the model with the given hyperparameters:

```
"""
define a seed for random number generator so the result will be reproducible
"""
import numpy as np
from tensorflow import random
np.random.seed(seed)
random.set_seed(seed)

from keras.models import Sequential
from keras.layers import Dense
# create model
model_1 = Sequential()
model_1.add(Dense(10, input_dim=X_train.shape[1], \
                  activation='relu'))
model_1.add(Dense(10, activation='relu'))
model_1.add(Dense(1))
# Compile model
model_1.compile(loss='mean_squared_error', optimizer='rmsprop')
```

3. Train the model on the training data with the given hyperparameters:

```
# train the model using training set while evaluating on test set
history=model_1.fit(X_train, y_train, batch_size = 50, \
                    epochs = 200, validation_data=(X_test, y_test), \
                    verbose=0)
```

4. Plot the trends for the **training error** and **test error**. Print the best accuracy that was reached for the training and validation set:

```
import matplotlib.pyplot as plt
import matplotlib
%matplotlib inline
matplotlib.rcParams['figure.figsize'] = (10.0, 8.0)
# plot training error and test error plots
plt.plot(history.history['loss'])
plt.plot(history.history['val_loss'])
plt.ylim((0, 25000))
```

```
plt.ylabel('loss')
plt.xlabel('epoch')
plt.legend(['train loss', 'validation loss'], loc='upper right')

# print the best accuracy reached on the test set
print("Lowest error on training set = ", \
      min(history.history['loss']))
print("Lowest error on validation set = ", \
      min(history.history['val_loss']))
```

The following is the expected output:

```
Lowest error on training set =  24.673954981565476
Lowest error on validation set =  25.11553382873535
```

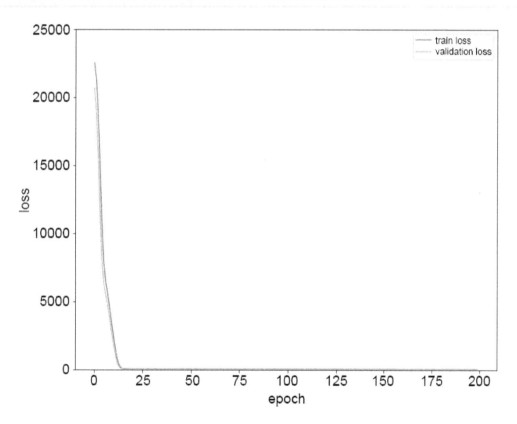

Figure 5.20: A plot of the training error and validation error during training
for the model without regularization

In the training error and validation error values, there is a very small gap between the training error and validation error, which is indicative of a low variance model, which is good.

5. Redefine the model by creating the same model architecture. However, this time, add a dropout regularization with **rate=0.1** to the first hidden layer of your model. Repeat *step 3* to train the model on the training data and repeat *step 4* to plot the trends for the training and validation errors. Then, print the best accuracy that was reached on the validation set:

```python
"""
define a seed for random number generator so the result will be
reproducible
"""
np.random.seed(seed)
random.set_seed(seed)

from keras.layers import Dropout
# create model
model_2 = Sequential()
model_2.add(Dense(10, input_dim=X_train.shape[1], \
                  activation='relu'))
model_2.add(Dropout(0.1))
model_2.add(Dense(10, activation='relu'))
model_2.add(Dense(1))
# Compile model
model_2.compile(loss='mean_squared_error', \
                optimizer='rmsprop')
# train the model using training set while evaluating on test set
history=model_2.fit(X_train, y_train, batch_size = 50, \
                    epochs = 200, validation_data=(X_test, y_test), \
                    verbose=0, shuffle=False)

matplotlib.rcParams['figure.figsize'] = (10.0, 8.0)
plt.plot(history.history['loss'])
plt.plot(history.history['val_loss'])
plt.ylim((0, 25000))
plt.ylabel('loss')
plt.xlabel('epoch')
plt.legend(['train loss', 'validation loss'], loc='upper right')
```

```
# print the best accuracy reached on the test set
print("Lowest error on training set = ", \
      min(history.history['loss']))
print("Lowest error on validation set = ", \
      min(history.history['val_loss']))
```

The following is the expected output:

```
Lowest error on training set =  407.8203821182251
Lowest error on validation set =  54.58488750457764
```

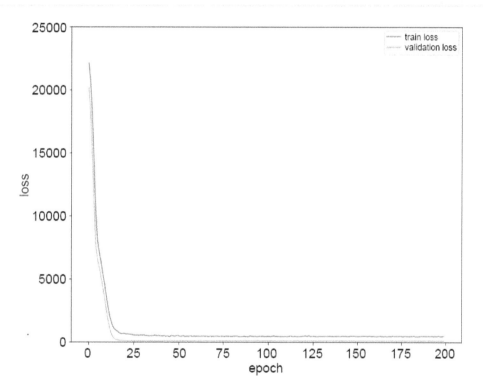

Figure 5.21: A plot of the training error and validation error during training for the model with dropout regularization (rate=0.1) in the first layer

There is a small gap between the training error and the validation error; however, the validation error is lower than the training error, indicating that the model is not overfitting the training data.

6. Repeat the previous step, this time adding dropout regularization with **rate=0.1** to both hidden layers of your model. Repeat *step 3* to train the model on the training data and repeat *step 4* to plot the trends for the training and validation errors. Then, print the best accuracy that was reached on the validation set:

```
"""
define a seed for random number generator so the result will be
reproducible
"""
np.random.seed(seed)
random.set_seed(seed)

# create model
model_3 = Sequential()
model_3.add(Dense(10, input_dim=X_train.shape[1], \
                  activation='relu'))
model_3.add(Dropout(0.1))
model_3.add(Dense(10, activation='relu'))
model_3.add(Dropout(0.1))
model_3.add(Dense(1))
# Compile model
model_3.compile(loss='mean_squared_error', \
                optimizer='rmsprop')
# train the model using training set while evaluating on test set
history=model_3.fit(X_train, y_train, batch_size = 50, \
                    epochs = 200, validation_data=(X_test, y_test), \
                    verbose=0, shuffle=False)

matplotlib.rcParams['figure.figsize'] = (10.0, 8.0)
plt.plot(history.history['loss'])
plt.plot(history.history['val_loss'])
plt.ylim((0, 25000))
plt.ylabel('loss')
plt.xlabel('epoch')
plt.legend(['train loss', 'validation loss'], loc='upper right')
```

```
# print the best accuracy reached on the test set
print("Lowest error on training set = ", \
      min(history.history['loss']))
print("Lowest error on validation set = ", \
      min(history.history['val_loss']))
```

The following is the expected output:

```
Lowest error on training set =  475.9299939632416
Lowest error on validation set =  61.646054649353026
```

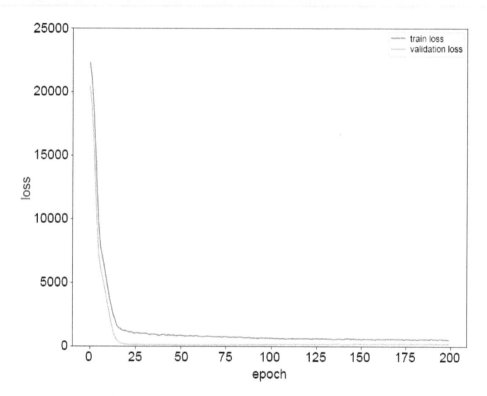

Figure 5.22: A plot of the training error and validation error during training for
the model with dropout regularization (rate=0.1) in both layers

The gap between the training error and validation error is slightly higher here,
mostly due to the increase in the training error as a result of the additional
regularization on the second hidden layer of the model.

7. Repeat the previous step, this time adding dropout regularization with **rate=0.2** in the first layer and **rate=0.1** in the second layer of your model. Repeat *step 3* to train the model on the training data and repeat *step 4* to plot the trends for the training and validation errors. Then, print the best accuracy that was reached on the validation set:

```
"""
define a seed for random number generator so the result will be
reproducible
"""
np.random.seed(seed)
random.set_seed(seed)

# create model
model_4 = Sequential()
model_4.add(Dense(10, input_dim=X_train.shape[1], \
                   activation='relu'))
model_4.add(Dropout(0.2))
model_4.add(Dense(10, activation='relu'))
model_4.add(Dropout(0.1))
model_4.add(Dense(1))
# Compile model
model_4.compile(loss='mean_squared_error', optimizer='rmsprop')
# train the model using training set while evaluating on test set
history=model_4.fit(X_train, y_train, batch_size = 50, epochs = 200, \
                    validation_data=(X_test, y_test), verbose=0)

matplotlib.rcParams['figure.figsize'] = (10.0, 8.0)
plt.plot(history.history['loss'])
plt.plot(history.history['val_loss'])
plt.ylim((0, 25000))
plt.ylabel('loss')
plt.xlabel('epoch')
plt.legend(['train loss', 'validation loss'], loc='upper right')

# print the best accuracy reached on the test set
print("Lowest error on training set = ", \
      min(history.history['loss']))
print("Lowest error on validation set = ", \
      min(history.history['val_loss']))
```

The following is the expected output:

```
Lowest error on training set =  935.1562484741211
Lowest error on validation set =  132.39965686798095
```

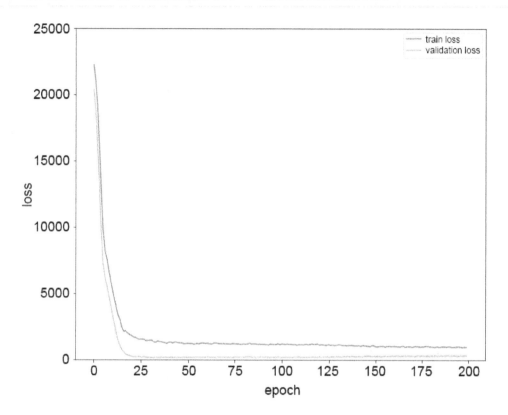

Figure 5.23: A plot of training errors and validation errors while training the model with dropout regularization, with rate=0.2 in the first layer and rate 0.1 in the second layer

The gap between the training error and validation error is slightly larger due to the increase in regularization. In this case, there was no overfitting in the original model. As a result, regularization increased the error rate on the training and validation dataset.

> **NOTE**
>
> To access the source code for this specific section, please refer to https://packt.live/38mtDo7.
>
> You can also run this example online at https://packt.live/31lsdmu.

ACTIVITY 5.03: HYPERPARAMETER TUNING ON THE AVILA PATTERN CLASSIFIER

In this activity, you will build a Keras model similar to those in the previous activities, but this time, you will add regularization methods to your model as well. Then, you will use scikit-learn optimizers to perform tuning on the model hyperparameters, including the hyperparameters of the regularizers. Follow these steps to complete this activity:

1. Load the dataset and import the libraries:

```
# Load The dataset
import pandas as pd
X = pd.read_csv('../data/avila-tr_feats.csv')
y = pd.read_csv('../data/avila-tr_target.csv')
```

2. Define a function that returns a Keras model with three hidden layers, the first of **size 10**, the second of **size 6**, and the third of **size 4**, and apply **L2 weight regularization** and a **ReLU activation** function on each hidden layer. Compile the model with the given parameters and return it from the model:

```
# Create the function that returns the keras model
from keras.models import Sequential
from keras.layers import Dense
from keras.regularizers import l2
def build_model(lambda_parameter):
    model = Sequential()
    model.add(Dense(10, input_dim=X.shape[1], \
                    activation='relu', \
                    kernel_regularizer=l2(lambda_parameter)))
    model.add(Dense(6, activation='relu', \
                    kernel_regularizer=l2(lambda_parameter)))
    model.add(Dense(4, activation='relu', \
                    kernel_regularizer=l2(lambda_parameter)))
```

```
    model.add(Dense(1, activation='sigmoid'))
    model.compile(loss='binary_crossentropy', \
                  optimizer='sgd', metrics=['accuracy'])
    return model
```

3. Set a seed, use a scikit-learn wrapper to wrap the model that we created in the previous step, and define the hyperparameters to scan. Finally, perform **GridSearchCV()** on the model using the hyperparameter's grid and fit the model:

```
from keras.wrappers.scikit_learn import KerasClassifier
from sklearn.model_selection import GridSearchCV
"""
define a seed for random number generator so the result will be
reproducible
"""
import numpy as np
from tensorflow import random
seed = 1
np.random.seed(seed)
random.set_seed(seed)
# create the Keras wrapper with scikit learn
model = KerasClassifier(build_fn=build_model, verbose=0, \
                        shuffle=False)
# define all the possible values for each hyperparameter
lambda_parameter = [0.01, 0.5, 1]
epochs = [50, 100]
batch_size = [20]
"""
create the dictionary containing all possible values of
hyperparameters
"""
param_grid = dict(lambda_parameter=lambda_parameter, \
                  epochs=epochs, batch_size=batch_size)
# perform 5-fold cross-validation for ??????? store the results
grid_seach = GridSearchCV(estimator=model, \
                          param_grid=param_grid, cv=5)
results_1 = grid_seach.fit(X, y)
```

4. Print the results for the best cross-validation score that's stored within the variable we created in the fit process. Iterate through all the parameters and print the mean of the accuracy across all the folds, the standard deviation of the accuracy, and the parameters themselves:

```
print("Best cross-validation score =", results_1.best_score_)
print("Parameters for Best cross-validation score=", \
      results_1.best_params_)

# print the results for all evaluated hyperparameter combinations
accuracy_means = results_1.cv_results_['mean_test_score']
accuracy_stds = results_1.cv_results_['std_test_score']
parameters = results_1.cv_results_['params']
for p in range(len(parameters)):
    print("Accuracy %f (std %f) for params %r" % \
            (accuracy_means[p], accuracy_stds[p], parameters[p]))
```

The following is the expected output:

```
Best cross-validation score = 0.7673058390617371
Parameters for Best cross-validation score= {'batch_size': 20,
'epochs': 100, 'lambda_parameter': 0.01}
Accuracy 0.764621 (std 0.004330) for params {'batch_size': 20,
'epochs': 50, 'lambda_parameter': 0.01}
Accuracy 0.589070 (std 0.008244) for params {'batch_size': 20,
'epochs': 50, 'lambda_parameter': 0.5}
Accuracy 0.589070 (std 0.008244) for params {'batch_size': 20,
'epochs': 50, 'lambda_parameter': 1}
Accuracy 0.767306 (std 0.015872) for params {'batch_size': 20,
'epochs': 100, 'lambda_parameter': 0.01}
Accuracy 0.589070 (std 0.008244) for params {'batch_size': 20,
'epochs': 100, 'lambda_parameter': 0.5}
Accuracy 0.589070 (std 0.008244) for params {'batch_size': 20,
'epochs': 100, 'lambda_parameter': 1}
```

5. Repeat *step 3* using **GridSearchCV()**, **lambda_parameter = [0.001, 0.01, 0.05, 0.1]**, **batch_size = [20]**, and **epochs = [100]**. Fit the model to the training data using **5-fold cross-validation** and print the results for the entire grid:

```
"""
define a seed for random number generator so the result will be
reproducible
```

```
"""
np.random.seed(seed)
random.set_seed(seed)
# create the Keras wrapper with scikit learn
model = KerasClassifier(build_fn=build_model, verbose=0, shuffle=False)
# define all the possible values for each hyperparameter
lambda_parameter = [0.001, 0.01, 0.05, 0.1]
epochs = [100]
batch_size = [20]
"""
create the dictionary containing all possible values of
hyperparameters
"""
param_grid = dict(lambda_parameter=lambda_parameter, \
                  epochs=epochs, batch_size=batch_size)
"""
search the grid, perform 5-fold cross-validation for each possible
combination, store the results
"""
grid_seach = GridSearchCV(estimator=model, \
                          param_grid=param_grid, cv=5)
results_2 = grid_seach.fit(X, y)

# print the results for best cross-validation score
print("Best cross-validation score =", results_2.best_score_)
print("Parameters for Best cross-validation score =", \
      results_2.best_params_)

# print the results for the entire grid
accuracy_means = results_2.cv_results_['mean_test_score']
accuracy_stds = results_2.cv_results_['std_test_score']
parameters = results_2.cv_results_['params']
for p in range(len(parameters)):
    print("Accuracy %f (std %f) for params %r" % \
          (accuracy_means[p], accuracy_stds[p], parameters[p]))
```

The following is the expected output:

```
Best cross-validation score = 0.786385428905487
Parameters for Best cross-validation score = {'batch_size': 20,
'epochs': 100, 'lambda_parameter': 0.001}
Accuracy 0.786385 (std 0.010177) for params {'batch_size': 20,
'epochs': 100, 'lambda_parameter': 0.001}
Accuracy 0.693960 (std 0.084994) for params {'batch_size': 20,
'epochs': 100, 'lambda_parameter': 0.01}
Accuracy 0.589070 (std 0.008244) for params {'batch_size': 20,
'epochs': 100, 'lambda_parameter': 0.05}
Accuracy 0.589070 (std 0.008244) for params {'batch_size': 20,
'epochs': 100, 'lambda_parameter': 0.1}
```

6. Redefine a function that returns a Keras model with three hidden layers, the first of **size 10**, the second of **size 6**, and the third of **size 4**, and apply **dropout regularization** and a **ReLU activation** function on each hidden layer. Compile the model with the given parameters and return it from the function:

```
# Create the function that returns the keras model
from keras.layers import Dropout
def build_model(rate):
    model = Sequential()
    model.add(Dense(10, input_dim=X.shape[1], activation='relu'))
    model.add(Dropout(rate))
    model.add(Dense(6, activation='relu'))
    model.add(Dropout(rate))
    model.add(Dense(4, activation='relu'))
    model.add(Dropout(rate))
    model.add(Dense(1, activation='sigmoid'))
    model.compile(loss='binary_crossentropy', \
                  optimizer='sgd', metrics=['accuracy'])
    return model
```

7. Use **rate = [0, 0.1, 0.2]** and **epochs = [50, 100]** and perform **GridSearchCV()** on the model. Fit the model to the training data using **5-fold cross-validation** and print the results for the entire grid:

```
"""
define a seed for random number generator so the result will be
reproducible
"""
```

```python
np.random.seed(seed)
random.set_seed(seed)
# create the Keras wrapper with scikit learn
model = KerasClassifier(build_fn=build_model, verbose=0,shuffle=False)
# define all the possible values for each hyperparameter
rate = [0, 0.1, 0.2]
epochs = [50, 100]
batch_size = [20]
"""
create the dictionary containing all possible values of
hyperparameters
"""


param_grid = dict(rate=rate, epochs=epochs, batch_size=batch_size)
"""
perform 5-fold cross-validation for 10 randomly selected
combinations, store the results
"""
grid_seach = GridSearchCV(estimator=model, \
                          param_grid=param_grid, cv=5)
results_3 = grid_seach.fit(X, y)


# print the results for best cross-validation score
print("Best cross-validation score =", results_3.best_score_)
print("Parameters for Best cross-validation score =", \
      results_3.best_params_)


# print the results for the entire grid
accuracy_means = results_3.cv_results_['mean_test_score']
accuracy_stds = results_3.cv_results_['std_test_score']
parameters = results_3.cv_results_['params']
for p in range(len(parameters)):
    print("Accuracy %f (std %f) for params %r" % \
          (accuracy_means[p], accuracy_stds[p], parameters[p]))
```

The following is the expected output:

```
Best cross-validation score= 0.7918504476547241
Parameters for Best cross-validation score= {'batch_size': 20,
'epochs': 100, 'rate': 0}
Accuracy 0.786769 (std 0.008255) for params {'batch_size': 20,
'epochs': 50, 'rate': 0}
Accuracy 0.764717 (std 0.007691) for params {'batch_size': 20,
'epochs': 50, 'rate': 0.1}
Accuracy 0.752637 (std 0.013546) for params {'batch_size': 20,
'epochs': 50, 'rate': 0.2}
Accuracy 0.791850 (std 0.008519) for params {'batch_size': 20,
'epochs': 100, 'rate': 0}
Accuracy 0.779291 (std 0.009504) for params {'batch_size': 20,
'epochs': 100, 'rate': 0.1}
Accuracy 0.767306 (std 0.005773) for params {'batch_size': 20,
'epochs': 100, 'rate': 0.2}
```

8. Repeat *step 5* using **rate = [0.0, 0.05, 0.1]** and **epochs = [100]**. Fit the model to the training data using **5-fold cross-validation** and print the results for the entire grid:

```
"""
define a seed for random number generator so the result will be
reproducible
"""
np.random.seed(seed)
random.set_seed(seed)
# create the Keras wrapper with scikit learn
model = KerasClassifier(build_fn=build_model, verbose=0, shuffle=False)
# define all the possible values for each hyperparameter
rate = [0.0, 0.05, 0.1]
epochs = [100]
batch_size = [20]
"""
create the dictionary containing all possible values of
hyperparameters
"""
param_grid = dict(rate=rate, epochs=epochs, batch_size=batch_size)
```

```
"""
perform 5-fold cross-validation for 10 randomly selected
combinations, store the results
"""

grid_seach = GridSearchCV(estimator=model, \
                          param_grid=param_grid, cv=5)
results_4 = grid_seach.fit(X, y)

# print the results for best cross-validation score
print("Best cross-validation score =", results_4.best_score_)
print("Parameters for Best cross-validation score =", \
      results_4.best_params_)

# print the results for the entire grid
accuracy_means = results_4.cv_results_['mean_test_score']
accuracy_stds = results_4.cv_results_['std_test_score']
parameters = results_4.cv_results_['params']
for p in range(len(parameters)):
    print("Accuracy %f (std %f) for params %r" % \
          (accuracy_means[p], accuracy_stds[p], parameters[p]))
```

The following is the expected output:

```
Best cross-validation score= 0.7862895488739013
Parameters for Best cross-validation score= {'batch_size': 20,
'epochs': 100, 'rate': 0.0}
Accuracy 0.786290 (std 0.013557) for params {'batch_size': 20,
'epochs': 100, 'rate': 0.0}
Accuracy 0.786098 (std 0.005184) for params {'batch_size': 20,
'epochs': 100, 'rate': 0.05}
Accuracy 0.772004 (std 0.013733) for params {'batch_size': 20,
'epochs': 100, 'rate': 0.1}
```

NOTE

To access the source code for this specific section, please refer to
https://packt.live/2D7HN0L.

This section does not currently have an online interactive example and will
need to be run locally.

CHAPTER 6: MODEL EVALUATION

ACTIVITY 6.01: COMPUTING THE ACCURACY AND NULL ACCURACY OF A NEURAL NETWORK WHEN WE CHANGE THE TRAIN/TEST SPLIT

In this activity, we will see that our **null accuracy** and **accuracy** will be affected by changing the **train/test** split. To implement this, the part of the code where the train/test split was defined has to be changed. We will use the same dataset that we used in *Exercise 6.02*, *Computing Accuracy and Null Accuracy with APS Failure for Scania Trucks Data*. Follow these steps to complete this activity:

1. Import the required libraries. Load the dataset using the pandas **read_csv** function and look at the first **five** rows of the dataset:

```
# Import the libraries
import numpy as np
import pandas as pd


# Load the Data
X = pd.read_csv("../data/aps_failure_training_feats.csv")
y = pd.read_csv("../data/aps_failure_training_target.csv")


# Use the head function to get a glimpse data
X.head()
```

The following table shows the output of the preceding code:

	aa_000	ab_000	ac_000	ad_000	ae_000	af_000	ag_000	ag_001	ag_002	ag_003	...	ee_002	ee_003	ee_004	ee_005	ee_006	ee_007
0	76698	0.0	2.130706e+09	280.0	0.0	0.0	0.0	0.0	0.0	0.0	...	1240520.0	493384.0	721044.0	469792.0	339156.0	157956.0
1	33058	0.0	0.000000e+00	0.0	0.0	0.0	0.0	0.0	0.0	0.0	...	421400.0	178064.0	293306.0	245416.0	133654.0	81140.0
2	41040	0.0	2.280000e+02	100.0	0.0	0.0	0.0	0.0	0.0	0.0	...	277378.0	159812.0	423992.0	409564.0	320746.0	158022.0
3	12	0.0	7.000000e+01	66.0	0.0	10.0	0.0	0.0	0.0	318.0	...	240.0	46.0	58.0	44.0	10.0	0.0
4	60874	0.0	1.368000e+03	458.0	0.0	0.0	0.0	0.0	0.0	0.0	...	622012.0	229790.0	405298.0	347188.0	286954.0	311560.0

5 rows × 170 columns

Figure 6.13: Initial five rows of the dataset

2. Change the **test_size** and **random_state** from **0.20** to **0.3** and **42** to **13**, respectively:

```
# Split the data into training and testing sets
from sklearn.model_selection import train_test_split
seed = 13
```

```
X_train, X_test, y_train, y_test = \
train_test_split(X, y, test_size=0.3, random_state=seed)
```

> **NOTE**
>
> If you use a different **random_state**, you may get a different **train/
> test** split, which may yield slightly different final results.

3. Scale the data using the **StandardScaler** function and use the scaler to scale the test data. Convert both into pandas DataFrames:

```
# Initialize StandardScaler
from sklearn.preprocessing import StandardScaler
sc = StandardScaler()

# Transform the training data
X_train = sc.fit_transform(X_train)
X_train = pd.DataFrame(X_train, columns=X_test.columns)

# Transform the testing data
X_test = sc.transform(X_test)
X_test = pd.DataFrame(X_test, columns = X_train.columns)
```

> **NOTE**
>
> The **sc.fit_transform()** function transforms the data, and the data is
> also converted into a **NumPy** array. We may need the data later for analysis
> as a DataFrame object, so the **pd.DataFrame()** function reconverts
> data into a DataFrame.

4. Import the libraries that are required to build a neural network architecture:

```
# Import the relevant Keras libraries
from keras.models import Sequential
from keras.layers import Dense
from keras.layers import Dropout
from tensorflow import random
```

5. Initiate the **Sequential** class:

```
# Initiate the Model with Sequential Class
np.random.seed(seed)
random.set_seed(seed)
model = Sequential()
```

6. Add five **Dense** layers to the network with **Dropout**. Set the first hidden layer so that it has a size of **64** with a dropout rate of **0.5**, the second hidden layer so that it has a size of **32** with a dropout rate of **0.4**, the third hidden layer so that it has a size of **16** with a dropout rate of **0.3**, the fourth hidden layer so that it has a size of **8** with a dropout rate of **0.2**, and the final hidden layer so that it has a size of **4** with a dropout rate of **0.1**. Set all the activation functions to **ReLU**:

```
# Add the hidden dense layers and with dropout Layer
model.add(Dense(units=64, activation='relu', \
                kernel_initializer='uniform', \
                input_dim=X_train.shape[1]))
model.add(Dropout(rate=0.5))
model.add(Dense(units=32, activation='relu', \
                kernel_initializer='uniform', \
                input_dim=X_train.shape[1]))
model.add(Dropout(rate=0.4))
model.add(Dense(units=16, activation='relu', \
                kernel_initializer='uniform', \
                input_dim=X_train.shape[1]))
model.add(Dropout(rate=0.3))
model.add(Dense(units=8, activation='relu', \
                kernel_initializer='uniform', \
                input_dim=X_train.shape[1]))
model.add(Dropout(rate=0.2))
model.add(Dense(units=4, activation='relu', \
                kernel_initializer='uniform'))
model.add(Dropout(rate=0.1))
```

7. Add an output **Dense** layer with a **sigmoid** activation function:

```
# Add Output Dense Layer
model.add(Dense(units=1, activation='sigmoid', \
                kernel_initializer='uniform'))
```

> **NOTE**
>
> Since the output is binary, we are using the **sigmoid** function. If the output is multiclass (that is, more than two classes), then the **softmax** function should be used.

8. Compile the network and fit the model. The metric that's being used here is **accuracy**:

```
# Compile the Model
model.compile(optimizer='adam', loss='binary_crossentropy', \
              metrics=['accuracy'])
```

> **NOTE**
>
> The metric name, which in our case is **accuracy**, is defined in the preceding code.

9. Fit the model with **100** epochs, a batch size of **20**, and a validation split of **0.2**:

```
# Fit the Model
model.fit(X_train, y_train, epochs=100, batch_size=20, \
          verbose=1, validation_split=0.2, shuffle=False)
```

10. Evaluate the model on the test dataset and print out the values for the **loss** and **accuracy**:

```
test_loss, test_acc = model.evaluate(X_test, y_test)
print(f'The loss on the test set is {test_loss:.4f} and \
the accuracy is {test_acc*100:.4f}%')
```

The preceding code produces the following output:

```
18000/18000 [==============================] - 0s 19us/step
The loss on the test set is 0.0766 and the accuracy is 98.9833%
```

The model returns an accuracy of **98.9833%**. But is it good enough? We can only get the answer to this question by comparing it against the null accuracy.

11. Now, compute the null accuracy. The **null accuracy** can be calculated using the **value_count** function of the **pandas** library, which we used in *Exercise 6.01, Calculating Null Accuracy on a Pacific Hurricanes Dataset*, of this chapter:

```
# Use the value_count function to calculate distinct class values
y_test['class'].value_counts()
```

The preceding code produces the following output:

```
0    17700
1      300
Name: class, dtype: int64
```

12. Calculate the **null accuracy**:

```
# Calculate the null accuracy
y_test['class'].value_counts(normalize=True).loc[0]
```

The preceding code produces the following output:

```
0.9833333333333333
```

> **NOTE**
>
> To access the source code for this specific section, please refer to
> https://packt.live/3eY7y1E.
>
> You can also run this example online at https://packt.live/2BzBO4n.

ACTIVITY 6.02: CALCULATING THE ROC CURVE AND AUC SCORE

The **ROC curve** and **AUC score** is an effective way to easily evaluate the performance of a binary classifier. In this activity, we will plot the **ROC curve** and calculate the **AUC score** of a model. We will use the same dataset and train the same model that we used in *Exercise 6.03, Deriving and Computing Metrics Based on a Confusion Matrix*. Continue with the same APS failure data, plot the **ROC curve**, and compute the **AUC score** of the model. Follow these steps to complete this activity:

1. Import the necessary libraries and load the data using the pandas **read_csv** function:

```
# Import the libraries
import numpy as np
import pandas as pd
```

```
# Load the Data
X = pd.read_csv("../data/aps_failure_training_feats.csv")
y = pd.read_csv("../data/aps_failure_training_target.csv")
```

2. Split the data into training and test datasets using the **train_test_split** function:

```
from sklearn.model_selection import train_test_split
seed = 42
X_train, X_test, y_train, y_test = \
train_test_split(X, y, test_size=0.20, random_state=seed)
```

3. Scale the feature data so that it has a **mean** of **0** and a **standard deviation** of **1** using the **StandardScaler** function. Fit the scaler in the **training data** and apply it to the **test data**:

```
from sklearn.preprocessing import StandardScaler
sc = StandardScaler()

# Transform the training data
X_train = sc.fit_transform(X_train)
X_train = pd.DataFrame(X_train,columns=X_test.columns)

# Transform the testing data
X_test = sc.transform(X_test)
X_test = pd.DataFrame(X_test,columns=X_train.columns)
```

4. Import the Keras libraries that are required for creating the model. Instantiate a Keras model of the **Sequential** class and add five hidden layers to the model, including dropout for each layer. The first hidden layer should have a size of **64** and a dropout rate of **0.5**. The second hidden layer should have a size of **32** and a dropout rate of **0.4**. The third hidden layer should have a size of **16** and a dropout rate of **0.3**. The fourth hidden layer should have a size of **8** and a dropout rate of **0.2**. The final hidden layer should have a size of **4** and a dropout rate of **0.1**. All the hidden layers should have **ReLU activation** functions and set **kernel_initializer = 'uniform'**. Add a final output layer to the model with a sigmoid activation function. Compile the model by calculating the accuracy metric during the training process:

```
# Import the relevant Keras libraries
from keras.models import Sequential
from keras.layers import Dense
```

```
from keras.layers import Dropout
from tensorflow import random

np.random.seed(seed)
random.set_seed(seed)
model = Sequential()

# Add the hidden dense layers with dropout Layer
model.add(Dense(units=64, activation='relu', \
                kernel_initializer='uniform', \
                input_dim=X_train.shape[1]))
model.add(Dropout(rate=0.5))
model.add(Dense(units=32, activation='relu', \
                kernel_initializer='uniform'))
model.add(Dropout(rate=0.4))
model.add(Dense(units=16, activation='relu', \
                kernel_initializer='uniform'))
model.add(Dropout(rate=0.3))
model.add(Dense(units=8, activation='relu', \
            kernel_initializer='uniform'))
model.add(Dropout(rate=0.2))
model.add(Dense(units=4, activation='relu', \
                kernel_initializer='uniform'))
model.add(Dropout(rate=0.1))

# Add Output Dense Layer
model.add(Dense(units=1, activation='sigmoid', \
                kernel_initializer='uniform'))

# Compile the Model
model.compile(optimizer='adam', loss='binary_crossentropy', \
            metrics=['accuracy'])
```

5. Fit the model to the training data by training for **100** epochs with **batch_size=20** and with **validation_split=0.2**:

```
model.fit(X_train, y_train, epochs=100, batch_size=20, \
          verbose=1, validation_split=0.2, shuffle=False)
```

6. Once the model has finished fitting to the training data, create a variable that is the result of the model's prediction on the test data using the model's **predict_proba** methods:

```
y_pred_prob = model.predict_proba(X_test)
```

7. Import **roc_curve** from scikit-learn and run the following code:

```
from sklearn.metrics import roc_curve
fpr, tpr, thresholds = roc_curve(y_test, y_pred_prob)
```

fpr = False positive rate (1 - specificity)

tpr = True positive rate (sensitivity)

thresholds = The threshold value of **y_pred_prob**

8. Run the following code to plot the **ROC curve** using **matplotlib.pyplot**:

```
import matplotlib.pyplot as plt
plt.plot(fpr, tpr)
plt.title("ROC Curve for APS Failure")
plt.xlabel("False Positive rate (1-Specificity)")
plt.ylabel("True Positive rate (Sensitivity)")
plt.grid(True)
plt.show()
```

The following plot shows the output of the preceding code:

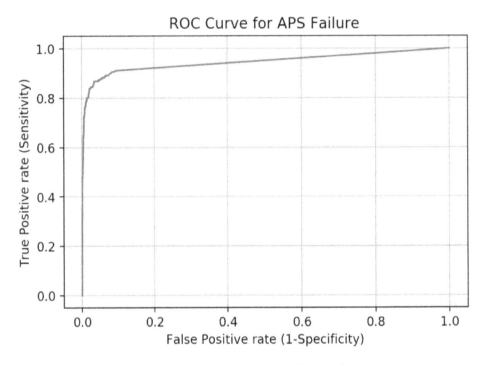

Figure 6.14: ROC curve of the APS failure dataset

9. Calculate the AUC score using the **roc_auc_score** function:

```
from sklearn.metrics import roc_auc_score
roc_auc_score(y_test,y_pred_prob)
```

The following is the output of the preceding code:

```
0.944787151628455
```

The AUC score of **94.4479%** suggests that our model is excellent, as per the general acceptable **AUC score** shown above.

> **NOTE**
>
> To access the source code for this specific section, please refer to https://packt.live/2NUOgyh.
>
> You can also run this example online at https://packt.live/2As33NH.

CHAPTER 7: COMPUTER VISION WITH CONVOLUTIONAL NEURAL NETWORKS

ACTIVITY 7.01: AMENDING OUR MODEL WITH MULTIPLE LAYERS AND THE USE OF SOFTMAX

Let's try and improve the performance of our image classification algorithm. There are many ways to improve its performance, and one of the most straightforward ways is by adding multiple ANN layers to the model, which we will learn about in this activity. We will also change the activation from sigmoid to softmax. Then, we can compare the result with that of the previous exercise. Follow these steps to complete this activity:

1. Import the **numpy** library and the necessary Keras libraries and classes:

    ```
    # Import the Libraries
    from keras.models import Sequential
    from keras.layers import Conv2D, MaxPool2D, Flatten, Dense
    import numpy as np
    from tensorflow import random
    ```

2. Now, initiate the model with the **Sequential** class:

    ```
    # Initiate the classifier
    seed = 1
    np.random.seed(seed)
    random.set_seed(seed)
    classifier=Sequential()
    ```

3. Add the first layer of the CNN, set the input shape to **(64, 64, 3)**, the dimension of each image, and the activation function as a ReLU. Then, add **32** feature detectors of size **(3, 3)**. Add two additional convolutional layers with **32** feature detectors of size **(3, 3)**, also with **ReLU activation** functions:

    ```
    classifier.add(Conv2D(32,(3,3),input_shape=(64,64,3),\
                    activation='relu'))
    classifier.add(Conv2D(32,(3,3),activation = 'relu'))
    classifier.add(Conv2D(32,(3,3),activation = 'relu'))
    ```

    32, (3, 3) means that there are **32** feature detectors of size **3x3**. As a good practice, always start with **32**; you can add **64** or **128** later.

4. Now, add the pooling layer with an image size of **2x2**:

    ```
    classifier.add(MaxPool2D(pool_size=(2,2)))
    ```

5. Flatten the output of the pooling layer by adding a flattening layer to the **CNN model**:

```
classifier.add(Flatten())
```

6. Add the first dense layer of the ANN. Here, **128** is the output of the number of nodes. As a good practice, **128** is good to get started. **activation** is **relu**. As a good practice, the power of two is preferred:

```
classifier.add(Dense(units=128,activation='relu'))
```

7. Add three more layers to the ANN of the same size, **128**, along with **ReLU activation** functions:

```
classifier.add(Dense(128,activation='relu'))
classifier.add(Dense(128,activation='relu'))
classifier.add(Dense(128,activation='relu'))
```

8. Add the output layer of the ANN. Replace the sigmoid function with **softmax**:

```
classifier.add(Dense(units=1,activation='softmax'))
```

9. Compile the network with an **Adam optimizer** and compute the accuracy during the training process:

```
# Compile The network
classifier.compile(optimizer='adam', loss='binary_crossentropy', \
                   metrics=['accuracy'])
```

10. Create training and test data generators. Rescale the training and test images by **1/255** so that all the values are between **0** and **1**. Set these parameters for the training data generators only – **shear_range=0.2**, **zoom_range=0.2**, and **horizontal_flip=True**:

```
from keras.preprocessing.image import ImageDataGenerator

train_datagen = ImageDataGenerator(rescale = 1./255, \
                                   shear_range = 0.2, \
                                   zoom_range = 0.2, \
                                   horizontal_flip = True)

test_datagen = ImageDataGenerator(rescale = 1./255)
```

11. Create a training set from the **training set** folder. **'../dataset/ training_set'** is the folder where our data has been placed. Our CNN model has an image size of **64x64**, so the same size should be passed here too. **batch_size** is the number of images in a single batch, which is **32**. **class_ mode** is set to **binary** since we are working on binary classifiers:

```
training_set = \
train_datagen.flow_from_directory('../dataset/training_set', \
                                  target_size = (64, 64), \
                                  batch_size = 32, \
                                  class_mode = 'binary')
```

12. Repeat *step 6* for the test by setting the folder to the location of the test images, that is, **'../dataset/test_set'**:

```
test_set = \
test_datagen.flow_from_directory('../dataset/test_set', \
                                 target_size = (64, 64), \
                                 batch_size = 32, \
                                 class_mode = 'binary')
```

13. Finally, fit the data. Set the **steps_per_epoch** to **10000** and the **validation_steps** to **2500**. The following step might take some time to execute:

```
classifier.fit_generator(training_set, steps_per_epoch = 10000, \
                         epochs = 2, validation_data = test_set, \
                         validation_steps = 2500, shuffle=False)
```

The preceding code produces the following output:

```
Epoch 1/2
10000/10000 [==============================] - 2452s 245ms/step -
loss: 8.1783 - accuracy: 0.4667 - val_loss: 11.4999 - val_accuracy:
0.4695
Epoch 2/2
10000/10000 [==============================] - 2496s 250ms/step -
loss: 8.1726 - accuracy: 0.4671 - val_loss: 10.5416 - val_accuracy:
0.4691
```

Note that the accuracy has decreased to **46.91%** due to the new softmax activation function.

> **NOTE**
>
> To access the source code for this specific section, please refer to
> https://packt.live/3gj0TiA.
>
> You can also run this example online at https://packt.live/2VIDj7e.

ACTIVITY 7.02: CLASSIFYING A NEW IMAGE

In this activity, you will try to classify another new image, just like we did in the preceding exercise. The image hasn't been exposed to the algorithm, so we will use this activity to test our algorithm. You can run any of the algorithms in this chapter (although the one that gets the highest accuracy is preferred) and then use the model to classify your images. Follow these steps to complete this activity:

1. Run one of the algorithms from this chapter.

2. Load the image and process it. **`test_image_2.jpg`** is the path of the test image. Change the path in the code where you have saved the dataset:

```
from keras.preprocessing import image
new_image = \
image.load_img('../test_image_2.jpg', target_size = (64, 64))
new_image
```

3. You can view the class labels using the following code:

```
training_set.class_indices
```

4. Process the image by converting it into a **numpy** array using the **img_to_array** function. Then, add an additional dimension along the 0th axis using numpy's **expand_dims** function:

```
new_image = image.img_to_array(new_image)
new_image = np.expand_dims(new_image, axis = 0)
```

5. Predict the new image by calling the **predict** method of the classifier:

```
result = classifier.predict(new_image)
```

6. Use the **class_indices** method with an **if**...**else** statement to map the 0 or 1 output of the prediction to a class label:

```
if result[0][0] == 1:
    prediction = 'It is a flower'
else:
    prediction = 'It is a car'
print(prediction)
```

The preceding code produces the following output:

```
It is a flower
```

test_image_2 is an image of a flower and was predicted to be a flower.

> **NOTE**
>
> To access the source code for this specific section, please refer to https://packt.live/38ny95E.
>
> You can also run this example online at https://packt.live/2VIM4Ow.

CHAPTER 8: TRANSFER LEARNING AND PRE-TRAINED MODELS

ACTIVITY 8.01: USING THE VGG16 NETWORK TO TRAIN A DEEP LEARNING NETWORK TO IDENTIFY IMAGES

Use the **VGG16** network to predict the image given (**test_image_1**). Before you start, ensure that you have downloaded the image (**test_image_1**) to your working directory. Follow these steps to complete this activity:

1. Import the **numpy** library and the necessary **Keras** libraries:

```
import numpy as np
from keras.applications.vgg16 import VGG16, preprocess_input
from keras.preprocessing import image
```

2. Initiate the model (note that, at this point, you can also view the architecture of the network, as shown in the following code):

```
classifier = VGG16()
classifier.summary()
```

classifier.summary() shows us the architecture of the network. The following points should be noted: it has a four-dimensional input shape (**None, 224, 224, 3**) and it has three convolutional layers.

The last four layers of the output are as follows:

```
flatten (Flatten)              (None, 25088)              0

fc1 (Dense)                    (None, 4096)               102764544

fc2 (Dense)                    (None, 4096)               16781312

predictions (Dense)            (None, 1000)               4097000
================================================================
Total params: 138,357,544
Trainable params: 138,357,544
Non-trainable params: 0

None
```

Figure 8.16: The architecture of the network

3. Load the image. `'../Data/Prediction/test_image_1.jpg'` is the path of the image on our system. It will be different on your system:

```
new_image = \
image.load_img('../Data/Prediction/test_image_1.jpg', \
               target_size=(224, 224))
new_image
```

The following figure shows the output of the preceding code:

Figure 8.17: The sample motorbike image

The target size should be **224x 224** since **VGG16** only accepts (**224,224**).

4. Change the image into an array by using the **img_to_array** function:

```
transformed_image = image.img_to_array(new_image)
transformed_image.shape
```

The preceding code provides the following output:

```
(224, 224, 3)
```

5. The image should be in a four-dimensional form for **VGG16** to allow further processing. Expand the dimension of the image, as follows:

```
transformed_image = np.expand_dims(transformed_image, axis=0)
transformed_image.shape
```

The preceding code provides the following output:

```
(1, 224, 224, 3)
```

6. Preprocess the image:

```
transformed_image = preprocess_input(transformed_image)
transformed_image
```

The following figure shows the output of the preceding code:

```
array([[[[-6.3939003e+01, -7.4778999e+01, -7.3680000e+01],
         [-2.1939003e+01, -3.5778999e+01, -3.8680000e+01],
         [-6.3939003e+01, -7.3778999e+01, -8.2680000e+01],
         ...,
         [-6.0939003e+01, -1.3778999e+01, -3.1680000e+01],
         [-7.0939003e+01, -2.2778999e+01, -4.3680000e+01],
         [-4.9939003e+01, -2.7789993e+00, -2.0680000e+01]],

        [[-2.4939003e+01, -3.3778999e+01, -3.9680000e+01],
         [-2.4939003e+01, -3.7778999e+01, -4.4680000e+01],
         [-2.9939003e+01, -3.9778999e+01, -4.8680000e+01],
         ...,
         [-6.7939003e+01, -1.8778999e+01, -3.7680000e+01],
         [-6.9939003e+01, -1.9778999e+01, -4.2680000e+01],
         [-7.7939003e+01, -2.8778999e+01, -4.9680000e+01]],
```

Figure 8.18: Image preprocessing

7. Create the **predictor** variable:

```
y_pred = classifier.predict(transformed_image)
y_pred
```

The following figure shows the output of the preceding code:

```
array([[[4.47333122e-07, 1.20946552e-07, 2.04147545e-06, 2.52621180e-06,
         6.90441425e-07, 7.73563841e-07, 2.69352967e-08, 9.62914100e-07,
         6.33308375e-08, 6.05552808e-09, 2.51603876e-08, 9.76482681e-08,
         9.47899537e-09, 4.40654730e-08, 4.79781761e-08, 1.10820743e-07,
         2.04076400e-07, 4.44985687e-07, 1.60248101e-06, 6.54645405e-08,
         9.36074329e-08, 2.09197353e-08, 1.36711648e-07, 6.79247250e-07,
         3.08072252e-08, 1.54558663e-07, 7.95182942e-09, 5.78766723e-09,
         1.74166360e-07, 1.18604691e-08, 7.84424614e-08, 2.26142323e-08,
         2.74102891e-08, 1.43111308e-07, 3.23035920e-06, 1.86695772e-07,
         5.57133092e-07, 3.49134872e-08, 1.87623090e-08, 1.51712968e-07,
         5.65604736e-08, 5.61646516e-08, 6.08605362e-08, 7.24316562e-09,
         2.91796027e-08, 1.61771148e-08, 4.54049882e-08, 1.36796743e-08,
         2.08325321e-08, 5.60503537e-08, 1.11806507e-07, 5.85347323e-07,
         6.61164279e-08, 3.45125919e-08, 4.91632761e-08, 4.12874961e-08,
         1.63845334e-06, 9.55561390e-08, 2.63248324e-07, 1.92033252e-08,
         2.32172539e-07, 2.96318120e-07, 1.38388089e-07, 1.73430777e-07,
         3.53732439e-08, 5.24874565e-07, 3.06662287e-08, 9.41523126e-08,
         6.36892707e-08, 5.16330516e-08, 1.57478812e-08, 8.49807691e-07,
         2.93759058e-07, 4.58372995e-07, 1.58948055e-07, 8.83325640e-07,
```

Figure 8.19: Creating the predictor variable

8. Check the shape of the image. It should be (**1**,**1000**). It's **1000** because, as we mentioned previously, the ImageNet database has **1000** categories of images. The predictor variable shows the probabilities of our image being one of those images:

```
y_pred.shape
```

The preceding code provides the following output:

```
(1, 1000)
```

9. Print the top five probabilities of what our image is using the **decode_predictions** function and pass the function of the predictor variable, **y_pred**, and the number of predictions and corresponding labels to output:

```
from keras.applications.vgg16 import decode_predictions
decode_predictions(y_pred, top=5)
```

The preceding code provides the following output:

```
[[('n03785016', 'moped', 0.8433369),
  ('n03791053', 'motor_scooter', 0.14188054),
  ('n03127747', 'crash_helmet', 0.007004856),
  ('n03208938', 'disk_brake', 0.0022349996),
  ('n04482393', 'tricycle', 0.0007717237)]]
```

The first column of the array is an internal code number. The second is the label, while the third is the probability of the image being the label.

10. Transform the predictions into a human-readable format. We need to extract the most probable label from the output, as follows:

```
label = decode_predictions(y_pred)
"""
Most likely result is retrieved, for example, the highest probability
"""
decoded_label = label[0][0]
# The classification is printed
print('%s (%.2f%%)' % (decoded_label[1], decoded_label[2]*100 ))
```

The preceding code provides the following output:

```
moped (84.33%)
```

Here, we can see that we have an **84.33%** probability that the picture is of a moped, which is close enough to a motorbike and probably represents the fact that motorbikes in the ImageNet dataset were labeled as mopeds.

> **NOTE**
>
> To access the source code for this specific section, please refer to https://packt.live/2C4nqRo.
>
> You can also run this example online at https://packt.live/31JMPL4.

ACTIVITY 8.02: IMAGE CLASSIFICATION WITH RESNET

In this activity, we will use another pre-trained network, known as **ResNet**. We have an image of television located at `../Data/Prediction/test_image_4`. We will use the **ResNet50** network to predict the image. Follow these steps to complete this activity:

1. Import the **numpy** library and the necessary **Keras** libraries:

```
import numpy as np
from keras.applications.resnet50 import ResNet50, preprocess_input
from keras.preprocessing import image
```

2. Initiate the ResNet50 model and print a summary of the model:

```
classifier = ResNet50()
classifier.summary()
```

classifier.summary() shows us the architecture of the network. The following points should be noted:

add_16 (Add)	(None, 7, 7, 2048)	0	bn5c_branch2c[0][0] activation_46[0][0]
activation_49 (Activation)	(None, 7, 7, 2048)	0	add_16[0][0]
avg_pool (GlobalAveragePooling2	(None, 2048)	0	activation_49[0][0]
fc1000 (Dense)	(None, 1000)	2049000	avg_pool[0][0]

```
Total params: 25,636,712
Trainable params: 25,583,592
Non-trainable params: 53,120
```

Figure 8.20: The last four layers of the output

NOTE

The last layer predictions (**Dense**) have **1000** values. This means that **VGG16** has a total of **1000** labels and that our image will be one of those **1000** labels.

3. Load the image. **'../Data/Prediction/test_image_4.jpg'** is the path of the image on our system. It will be different on your system:

```
new_image = \
image.load_img('../Data/Prediction/test_image_4.jpg', \
            target_size=(224, 224))
new_image
```

The following is the output of the preceding code:

Figure 8.21: A sample image of a television

The target size should be **224x224** since **ResNet50** only accepts (**224,224**).

4. Change the image into an array by using the **img_to_array** function:

```
transformed_image = image.img_to_array(new_image)
transformed_image.shape
```

5. The image has to be in a four-dimensional form for **ResNet50** to allow further processing. Expand the dimensions of the image along the 0th axis using the **expand_dims** function:

```
transformed_image = np.expand_dims(transformed_image, axis=0)
transformed_image.shape
```

6. Preprocess the image using the **preprocess_input** function:

```
transformed_image = preprocess_input(transformed_image)
transformed_image
```

7. Create the predictor variable by using the classifier to predict the image using it's **predict** method:

```
y_pred = classifier.predict(transformed_image)
y_pred
```

8. Check the shape of the image. It should be (**1 , 1000**):

```
y_pred.shape
```

The preceding code provides the following output:

```
(1, 1000)
```

9. Select the top five probabilities of what our image is using the **decode_predictions** function and by passing the predictor variable, **y_pred**, as the argument and the top number of predictions and corresponding labels:

```
from keras.applications.resnet50 import decode_predictions
decode_predictions(y_pred, top=5)
```

The preceding code provides the following output:

```
[[('n04404412', 'television', 0.99673873),
  ('n04372370', 'switch', 0.0009829825),
  ('n04152593', 'screen', 0.00095111143),
  ('n03782006', 'monitor', 0.0006477369),
  ('n04069434', 'reflex_camera', 8.5398955e-05)]]
```

The first column of the array is an internal code number. The second is the label, while the third is the probability of the image matching the label.

10. Put the predictions in a human-readable format. Print the most probable label from the output from the result of the **decode_predictions** function:

```
label = decode_predictions(y_pred)
"""
Most likely result is retrieved, for example,
the highest probability
"""
decoded_label = label[0][0]
# The classification is printed
print('%s (%.2f%%)' % (decoded_label[1], decoded_label[2]*100 ))
```

The preceding code produces the following output:

```
television (99.67%)
```

> **NOTE**
>
> To access the source code for this specific section, please refer to https://packt.live/38rEe0M.
>
> You can also run this example online at https://packt.live/2YV5xxo.

CHAPTER 9: SEQUENTIAL MODELING WITH RECURRENT NEURAL NETWORKS

ACTIVITY 9.01: PREDICTING THE TREND OF AMAZON'S STOCK PRICE USING AN LSTM WITH 50 UNITS (NEURONS)

In this activity, we will examine the stock price of Amazon for the last 5 years—from January 1, 2014, to December 31, 2018. In doing so, we will try to predict and forecast the company's future trend for January 2019 using an **RNN** and **LSTM**. We have the actual values for January 2019, so we can compare our predictions to the actual values later. Follow these steps to complete this activity:

1. Import the required libraries:

    ```
    import numpy as np
    import matplotlib.pyplot as plt
    import pandas as pd
    from tensorflow import random
    ```

2. Import the dataset using the pandas **read_csv** function and look at the first five rows of the dataset using the **head** method:

    ```
    dataset_training = pd.read_csv('../AMZN_train.csv')
    dataset_training.head()
    ```

    The following figure shows the output of the preceding code:

	Date	Open	High	Low	Close	Adj Close	Volume
0	2014-01-02	398.799988	399.359985	394.019989	397.970001	397.970001	2137800
1	2014-01-03	398.290009	402.709991	396.220001	396.440002	396.440002	2210200
2	2014-01-06	395.850006	397.000000	388.420013	393.630005	393.630005	3170600
3	2014-01-07	395.040009	398.470001	394.290009	398.029999	398.029999	1916000
4	2014-01-08	398.470001	403.000000	396.040009	401.920013	401.920013	2316500

Figure 9.24: The first five rows of the dataset

3. We are going to make our prediction using the **Open** stock price; therefore, select the **Open** stock price column from the dataset and print the values:

    ```
    training_data = dataset_training[['Open']].values
    training_data
    ```

The preceding code produces the following output:

```
array([[ 398.799988],
       [ 398.290009],
       [ 395.850006],
       ...,
       [1454.199951],
       [1473.349976],
       [1510.800049]])
```

4. Then, perform feature scaling by normalizing the data using **MinMaxScaler** and setting the range of the features so that they have a minimum value of zero and a maximum value of one. Use the **fit_transform** method of the scaler on the training data:

```
from sklearn.preprocessing import MinMaxScaler
sc = MinMaxScaler(feature_range = (0, 1))
training_data_scaled = sc.fit_transform(training_data)


training_data_scaled
```

The preceding code produces the following output:

```
array([[0.06523313],
       [0.06494233],
       [0.06355099],
       ...,
       [0.66704299],
       [0.67796271],
       [0.69931748]])
```

5. Create the data to get **60** timestamps from the current instance. We chose **60** here as it will give us a sufficient number of previous instances in order to understand the trend; technically, this can be any number, but **60** is the optimal value. Additionally, the upper bound value here is **1258**, which is the index or count of rows (or records) in the training set:

```
X_train = []
y_train = []
for i in range(60, 1258):
    X_train.append(training_data_scaled[i-60:i, 0])
    y_train.append(training_data_scaled[i, 0])
X_train, y_train = np.array(X_train), np.array(y_train)
```

6. Reshape the data to add an extra dimension to the end of **X_train** using NumPy's **reshape** function:

```
X_train = np.reshape(X_train, (X_train.shape[0], \
                      X_train.shape[1], 1))
```

7. Import the following libraries to build the RNN:

```
from keras.models import Sequential
from keras.layers import Dense, LSTM, Dropout
```

8. Set the seed and initiate the sequential model, as follows:

```
seed = 1
np.random.seed(seed)
random.set_seed(seed)
model = Sequential()
```

9. Add an **LSTM** layer to the network with **50** units, set the **return_sequences** argument to **True**, and set the **input_shape** argument to **(X_train. shape[1], 1)**. Add three additional **LSTM** layers, each with **50** units, and set the **return_sequences** argument to **True** for the first two. Add a final output layer of size 1:

```
model.add(LSTM(units = 50, return_sequences = True, \
          input_shape = (X_train.shape[1], 1)))

# Adding a second LSTM layer
model.add(LSTM(units = 50, return_sequences = True))

# Adding a third LSTM layer
model.add(LSTM(units = 50, return_sequences = True))

# Adding a fourth LSTM layer
model.add(LSTM(units = 50))

# Adding the output layer
model.add(Dense(units = 1))
```

10. Compile the network with an **adam** optimizer and use **Mean Squared Error** for the loss. Fit the model to the training data for **100** epochs with a batch size of **32**:

```
# Compiling the RNN
model.compile(optimizer = 'adam', loss = 'mean_squared_error')


# Fitting the RNN to the Training set
model.fit(X_train, y_train, epochs = 100, batch_size = 32)
```

11. Load and process the test data (which is treated as actual data here) and select the column representing the value of **Open** stock data:

```
dataset_testing = pd.read_csv('../AMZN_test.csv')
actual_stock_price = dataset_testing[['Open']].values
actual_stock_price
```

12. Concatenate the data since we will need **60** previous instances to get the stock price for each day. Therefore, we will need both the training and test data:

```
total_data = pd.concat((dataset_training['Open'], \
                        dataset_testing['Open']), axis = 0)
```

13. Reshape and scale the input to prepare the test data. Note that we are predicting the January monthly trend, which has **21** financial days, so in order to prepare the test set, we take the lower bound value as **60** and the upper bound value as **81**. This ensures that the difference of **21** is maintained:

```
inputs = total_data[len(total_data) \
         - len(dataset_testing) - 60:].values
inputs = inputs.reshape(-1,1)
inputs = sc.transform(inputs)
X_test = []
for i in range(60, 81):
    X_test.append(inputs[i-60:i, 0])
X_test = np.array(X_test)
X_test = np.reshape(X_test, (X_test.shape[0], \
                             X_test.shape[1], 1))
predicted_stock_price = model.predict(X_test)
predicted_stock_price = \
sc.inverse_transform(predicted_stock_price)
```

14. Visualize the results by plotting the actual stock price and plotting the predicted stock price:

```
# Visualizing the results
plt.plot(actual_stock_price, color = 'green', \
        label = 'Real Amazon Stock Price',ls='--')
plt.plot(predicted_stock_price, color = 'red', \
        label = 'Predicted Amazon Stock Price',ls='-')
plt.title('Predicted Stock Price')
plt.xlabel('Time in days')
plt.ylabel('Real Stock Price')
plt.legend()
plt.show()
```

Please note that your results may differ slightly from the actual stock price of Amazon.

Expected output:

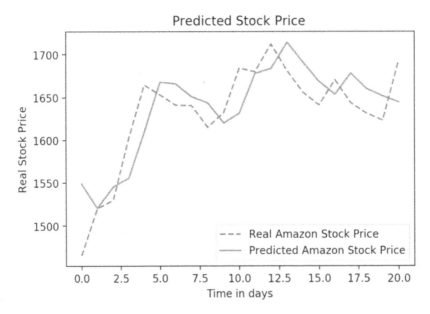

Figure 9.25: Real versus predicted stock prices

As shown in the preceding plot, the trends of the predicted and real prices are pretty much the same; the line has the same peaks and troughs. This is possible because of LSTM's ability to remember sequenced data. A traditional feedforward neural network would not have been able to forecast this result. This is the true power of **LSTM** and **RNNs**.

> **NOTE**
>
> To access the source code for this specific section, please refer to
> https://packt.live/3goQO3I.
>
> You can also run this example online at https://packt.live/2VIMq7O.

ACTIVITY 9.02: PREDICTING AMAZON'S STOCK PRICE WITH ADDED REGULARIZATION

In this activity, we will examine the stock price of Amazon over the last 5 years, from January 1, 2014, to December 31, 2018. In doing so, we will try to predict and forecast the company's future trend for January 2019 using RNNs and an LSTM. We have the actual values for January 2019, so we will be able to compare our predictions with the actual values later. Initially, we predicted the trend of Amazon's stock price using an LSTM with 50 units (or neurons). In this activity, we will also add dropout regularization and compare the results with *Activity 9.01, Predicting the Trend of Amazon's Stock Price Using an LSTM with 50 Units (Neurons)*. Follow these steps to complete this activity:

1. Import the required libraries:

```
import numpy as np
import matplotlib.pyplot as plt
import pandas as pd
from tensorflow import random
```

2. Import the dataset using the pandas **read_csv** function and look at the first five rows of the dataset using the **head** method:

```
dataset_training = pd.read_csv('../AMZN_train.csv')
dataset_training.head()
```

3. We are going to make our prediction using the **Open** stock price; therefore, select the **Open** stock price column from the dataset and print the values:

```
training_data = dataset_training[['Open']].values
training_data
```

The preceding code produces the following output:

```
array([[ 398.799988],
       [ 398.290009],
       [ 395.850006],
```

```
    ...,
    [1454.199951],
    [1473.349976],
    [1510.800049]])
```

4. Then, perform feature scaling by normalizing the data using **MinMaxScaler** and setting the range of the features so that they have a minimum value of **0** and a maximum value of one. Use the **fit_transform** method of the scaler on the training data:

```
from sklearn.preprocessing import MinMaxScaler
sc = MinMaxScaler(feature_range = (0, 1))
training_data_scaled = sc.fit_transform(training_data)

training_data_scaled
```

The preceding code produces the following output:

```
array([[0.06523313],
       [0.06494233],
       [0.06355099],
       ...,
       [0.66704299],
       [0.67796271],
       [0.69931748]])
```

5. Create the data to get **60** timestamps from the current instance. We chose **60** here as it will give us a sufficient number of previous instances in order to understand the trend; technically, this can be any number, but **60** is the optimal value. Additionally, the upper bound value here is **1258**, which is the index or count of rows (or records) in the training set:

```
X_train = []
y_train = []
for i in range(60, 1258):
    X_train.append(training_data_scaled[i-60:i, 0])
    y_train.append(training_data_scaled[i, 0])
X_train, y_train = np.array(X_train), np.array(y_train)
```

6. Reshape the data to add an extra dimension to the end of **X_train** using NumPy's **reshape** function:

```
X_train = np.reshape(X_train, (X_train.shape[0], \
                               X_train.shape[1], 1))
```

7. Import the following Keras libraries to build the RNN:

```
from keras.models import Sequential
from keras.layers import Dense, LSTM, Dropout
```

8. Set the seed and initiate the sequential model, as follows:

```
seed = 1
np.random.seed(seed)
random.set_seed(seed)
model = Sequential()
```

9. Add an LSTM layer to the network with 50 units, set the **return_sequences** argument to **True**, and set the **input_shape** argument to **(X_train. shape[1], 1)**. Add dropout to the model with **rate=0.2**. Add three additional LSTM layers, each with **50** units, and set the **return_sequences** argument to **True** for the first two. After each **LSTM** layer, add a dropout with **rate=0.2**. Add a final output layer of size **1**:

```
model.add(LSTM(units = 50, return_sequences = True, \
               input_shape = (X_train.shape[1], 1)))
model.add(Dropout(0.2))

# Adding a second LSTM layer and some Dropout regularization
model.add(LSTM(units = 50, return_sequences = True))
model.add(Dropout(0.2))

# Adding a third LSTM layer and some Dropout regularization
model.add(LSTM(units = 50, return_sequences = True))
model.add(Dropout(0.2))

# Adding a fourth LSTM layer and some Dropout regularization
model.add(LSTM(units = 50))
model.add(Dropout(0.2))

# Adding the output layer
model.add(Dense(units = 1))
```

10. Compile the network with an **adam** optimizer and use **Mean Squared Error** for the loss. Fit the model to the training data for **100** epochs with a batch size of **32**:

```
# Compiling the RNN
model.compile(optimizer = 'adam', loss = 'mean_squared_error')


# Fitting the RNN to the Training set
model.fit(X_train, y_train, epochs = 100, batch_size = 32)
```

11. Load and process the test data (which is treated as actual data here) and select the column representing the value of **Open** stock data:

```
dataset_testing = pd.read_csv('../AMZN_test.csv')
actual_stock_price = dataset_testing[['Open']].values
actual_stock_price
```

12. Concatenate the data since we will need **60** previous instances to get the stock price for each day. Therefore, we will need both the training and test data:

```
total_data = pd.concat((dataset_training['Open'], \
                        dataset_testing['Open']), axis = 0)
```

13. Reshape and scale the input to prepare the test data. Note that we are predicting the January monthly trend, which has **21** financial days, so in order to prepare the test set, we take the lower bound value as **60** and the upper bound value as **81**. This ensures that the difference of **21** is maintained:

```
inputs = total_data[len(total_data) \
         - len(dataset_testing) - 60:].values
inputs = inputs.reshape(-1,1)
inputs = sc.transform(inputs)
X_test = []
for i in range(60, 81):
    X_test.append(inputs[i-60:i, 0])
X_test = np.array(X_test)
X_test = np.reshape(X_test, (X_test.shape[0], \
                             X_test.shape[1], 1))
predicted_stock_price = model.predict(X_test)
predicted_stock_price = \
sc.inverse_transform(predicted_stock_price)
```

14. Visualize the results by plotting the actual stock price and plotting the predicted stock price:

```
# Visualizing the results
plt.plot(actual_stock_price, color = 'green', \
        label = 'Real Amazon Stock Price',ls='--')
plt.plot(predicted_stock_price, color = 'red', \
        label = 'Predicted Amazon Stock Price',ls='-')
plt.title('Predicted Stock Price')
plt.xlabel('Time in days')
plt.ylabel('Real Stock Price')
plt.legend()
plt.show()
```

Please note that your results may differ slightly to the actual stock price.

Expected output:

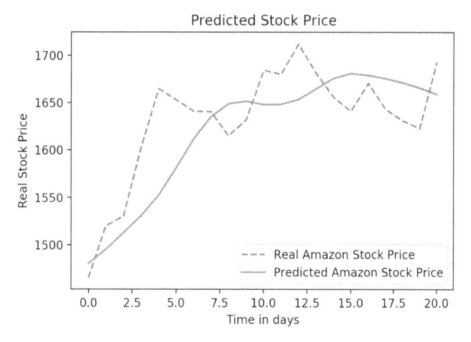

Figure 9.26: Real versus predicted stock prices

In the following figure, the first plot displays the predicted output of the model with regularization from Activity 9.02, and the second displays the predicted output without regularization from Activity 9.01. As you can see, adding dropout regularization does not fit the data as accurately. So, in this case, it is better not to use regularization, or to use dropout regularization with a lower dropout rate :

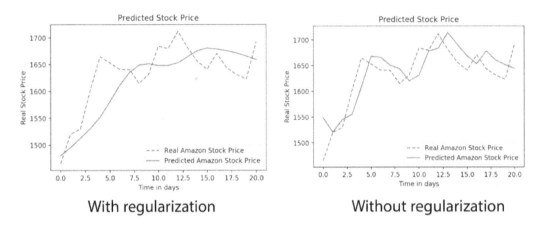

With regularization   Without regularization

Figure 9.27: Comparing the results of Activity 9.01 and Activity 9.02

> **NOTE**
>
> To access the source code for this specific section, please refer to https://packt.live/2YTpxR7.
>
> You can also run this example online at https://packt.live/3dY5Bku.

ACTIVITY 9.03: PREDICTING THE TREND OF AMAZON'S STOCK PRICE USING AN LSTM WITH AN INCREASING NUMBER OF LSTM NEURONS (100 UNITS)

In this activity, we will examine the stock price of Amazon over the last 5 years, from January 1, 2014, to December 31, 2018. We will try to predict and forecast the company's future trend for January 2019 using **RNNs** with four **LSTM** layers, each with **100** units. We have the actual values for January 2019, so we will be able to compare our predictions with the actual values later. You can also compare the output difference with *Activity 9.01, Predicting the Trend of Amazon's Stock Price Using an LSTM with 50 Units (Neurons)*. Follow these steps to complete this activity:

1. Import the required libraries:

```
import numpy as np
import matplotlib.pyplot as plt
import pandas as pd
from tensorflow import random
```

2. Import the dataset using the pandas **read_csv** function and look at the first five rows of the dataset using the **head** method:

```
dataset_training = pd.read_csv('../AMZN_train.csv')
dataset_training.head()
```

3. We are going to make our prediction using the **Open** stock price; therefore, select the **Open** stock price column from the dataset and print the values:

```
training_data = dataset_training[['Open']].values
training_data
```

4. Then, perform feature scaling by normalizing the data using **MinMaxScaler** and setting the range of the features so that they have a minimum value of zero and a maximum value of one. Use the **fit_transform** method of the scaler on the training data:

```
from sklearn.preprocessing import MinMaxScaler
sc = MinMaxScaler(feature_range = (0, 1))
training_data_scaled = sc.fit_transform(training_data)
training_data_scaled
```

5. Create the data to get **60** timestamps from the current instance. We chose **60** here as it will give us a sufficient number of previous instances in order to understand the trend; technically, this can be any number, but **60** is the optimal value. Additionally, the upper bound value here is **1258**, which is the index or count of rows (or records) in the training set:

```
X_train = []
y_train = []
for i in range(60, 1258):
    X_train.append(training_data_scaled[i-60:i, 0])
    y_train.append(training_data_scaled[i, 0])
X_train, y_train = np.array(X_train), np.array(y_train)
```

6. Reshape the data to add an extra dimension to the end of **X_train** using NumPy's **reshape** function:

```
X_train = np.reshape(X_train, (X_train.shape[0], \
                               X_train.shape[1], 1))
```

7. Import the following Keras libraries to build the RNN:

```
from keras.models import Sequential
from keras.layers import Dense, LSTM, Dropout
```

8. Set the seed and initiate the sequential model:

```
seed = 1
np.random.seed(seed)
random.set_seed(seed)
model = Sequential()
```

9. Add an LSTM layer to the network with **100** units, set the **return_sequences** argument to **True**, and set the **input_shape** argument to **(X_train.shape[1], 1)**. Add three additional **LSTM** layers, each with **100** units, and set the **return_sequences** argument to **True** for the first two. Add a final output layer of size **1**:

```
model.add(LSTM(units = 100, return_sequences = True, \
               input_shape = (X_train.shape[1], 1)))

# Adding a second LSTM layer
model.add(LSTM(units = 100, return_sequences = True))

# Adding a third LSTM layer
model.add(LSTM(units = 100, return_sequences = True))

# Adding a fourth LSTM layer
model.add(LSTM(units = 100))

# Adding the output layer
model.add(Dense(units = 1))
```

10. Compile the network with an **adam** optimizer and use **Mean Squared Error** for the loss. Fit the model to the training data for **100** epochs with a batch size of **32**:

```
# Compiling the RNN
model.compile(optimizer = 'adam', loss = 'mean_squared_error')


# Fitting the RNN to the Training set
model.fit(X_train, y_train, epochs = 100, batch_size = 32)
```

11. Load and process the test data (which is treated as actual data here) and select the column representing the value of open stock data:

```
dataset_testing = pd.read_csv('../AMZN_test.csv')
actual_stock_price = dataset_testing[['Open']].values
actual_stock_price
```

12. Concatenate the data since we will need **60** previous instances to get the stock price for each day. Therefore, we will need both the training and test data:

```
total_data = pd.concat((dataset_training['Open'], \
                        dataset_testing['Open']), axis = 0)
```

13. Reshape and scale the input to prepare the test data. Note that we are predicting the January monthly trend, which has **21** financial days, so in order to prepare the test set, we take the lower bound value as **60** and the upper bound value as **81**. This ensures that the difference of **21** is maintained:

```
inputs = total_data[len(total_data) \
         - len(dataset_testing) - 60:].values
inputs = inputs.reshape(-1,1)
inputs = sc.transform(inputs)
X_test = []
for i in range(60, 81):
    X_test.append(inputs[i-60:i, 0])
X_test = np.array(X_test)
X_test = np.reshape(X_test, (X_test.shape[0], \
                             X_test.shape[1], 1))
predicted_stock_price = model.predict(X_test)
predicted_stock_price = \
sc.inverse_transform(predicted_stock_price)
```

14. Visualize the results by plotting the actual stock price and plotting the predicted stock price:

```
plt.plot(actual_stock_price, color = 'green', \
        label = 'Actual Amazon Stock Price',ls='--')
plt.plot(predicted_stock_price, color = 'red', \
        label = 'Predicted Amazon Stock Price',ls='-')
plt.title('Predicted Stock Price')
plt.xlabel('Time in days')
plt.ylabel('Real Stock Price')
plt.legend()
plt.show()
```

Please note that your results may differ slightly from the actual stock price.

Expected output:

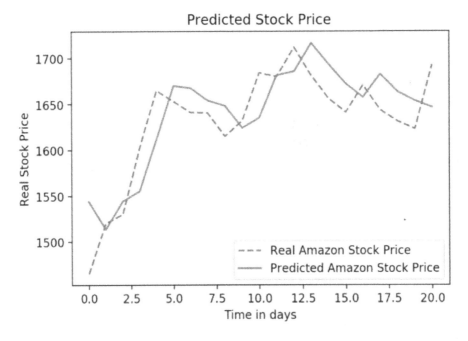

Figure 9.28: Real versus predicted stock prices

So, if we compare the results of the **LSTM** with **50** units (from *Activity 9.01, Predicting the Trend of Amazon's Stock Price Using an LSTM with 50 Units (Neurons)*) and the **LSTM** with **100** units in this activity, we get trends with **100** units. Also, note that when we run the **LSTM** with **100** units, it takes more computational time than the **LSTM** with **50** units. A trade-off needs to be considered in such cases:

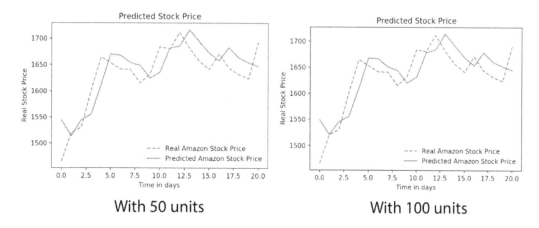

Figure 9.29: Comparing the real versus predicted stock price with 50 and 100 units

NOTE

To access the source code for this specific section, please refer to https://packt.live/31NQkQy.

You can also run this example online at https://packt.live/2ZCZ4GR.

INDEX

V

W

www.ingramcontent.com/pod-product-compliance
Lightning Source LLC
Chambersburg PA
CBHW081454050326
40690CB00015B/2797